Measurement of Soft Tissue Elasticity *in Vivo*

Techniques and Applications

Measurement of Soft Tissue Elasticity *in Vivo*

Techniques and Applications

Yong-Ping Zheng
Yan-Ping Huang

CRC Press
Taylor & Francis Group
Boca Raton London New York

CRC Press is an imprint of the
Taylor & Francis Group, an **informa** business

CRC Press
Taylor & Francis Group
6000 Broken Sound Parkway NW, Suite 300
Boca Raton, FL 33487-2742

First issued in paperback 2019

© 2016 by Taylor & Francis Group, LLC
CRC Press is an imprint of Taylor & Francis Group, an Informa business

No claim to original U.S. Government works

ISBN-13: 978-1-4665-7628-5 (hbk)
ISBN-13: 978-0-367-37720-5 (pbk)

Library of Congress Cataloging-in-Publication Data

Huang, Yan-Ping, 1980- , author.
    Measurement of soft tissue elasticity in vivo : techniques and applications / Yan-Ping Huang and Yong-Ping Zheng.
        p. ; cm.
    Includes bibliographical references and index.
    ISBN 978-1-4665-7628-5 (alk. paper)
    I. Zheng, Yong-Ping, 1966- , author. II. Title.
    [DNLM: 1. Elasticity Imaging Techniques--methods. 2. Tissues--physiology. WN 208]

RC78.7.U4
616.07'543--dc23   2015027182

**Visit the Taylor & Francis Web site at
http://www.taylorandfrancis.com**

**and the CRC Press Web site at
http://www.crcpress.com**

Contents

Preface

A N IMPORTANT MOTIVATION FOR US TO WRITE THIS BOOK was the lack of a standardised method for tissue elasticity measurement *in vivo*, in spite of so many different techniques available. These techniques have been developed based on different physical principles mainly over the past 30 years, although the first indentation device for the assessment of skin *in vivo* can be traced back to 1912. Researchers working on technological developments tend to focus on the techniques they initiated and make efforts to find more applications. On the other hand, clinicians or researchers working on the applications may either stick to a technique familiar to them or face a challenge of how to select a suitable technique for their applications. For researchers working on mechanics, it is well known that only two independent mechanical parameters are required to describe a homogenous, isotropic, linear and elastic material, which can be any two of Young's modulus, shear modulus and Poisson's ratio. Every material possesses intrinsic mechanical parameters. However, when different techniques are developed by different research groups based on different physical principles for measuring mechanical properties of different soft tissues, highly different parameters are used to represent tissue elasticity. One aim of this book is to provide both engineers interested in developing new methods and clinicians using techniques with an overview of the existing *in vivo* tissue elasticity measurement methods, their physical principles, advantages, disadvantages and assumptions.

Since Prof. Y.C. Fung published his classic textbook on biomechanics, *Biomechanics: Mechanical Properties of Living Tissues*, in 1993, there have been limited techniques available for the measurement of soft tissue elasticity *in vivo*. If readers aim to form a systematic understanding of the fundamental biomechanics of different body tissues, this is the right book to read. The entire field of ultrasound-assisted tissue elasticity measurement and imaging techniques has been rapidly developing since the early 1990s with the representative paper of Prof. J. Ophir and coworkers, 'Elastography: a quantitative method for imaging the elasticity of biological tissues'. *Ultrasound Imaging* 13(2):111–134, in 1991. Later, magnetic resonance imaging (MRI) and optical imaging-assisted techniques were developed. Since then, researchers all over the world have proposed many innovative methods aiming to provide more accurate, intrinsic and convenient measurement of tissue elasticity *in vivo*, including sonoelastography, transient elastography (TE), intravascular elastography, vibro-acoustography, supersonic shear imaging (SSI), harmonic motion imaging, acoustic radiation force impulse imaging (ARFI), magnetic resonance elastography (MRE), optical elastography, etc. More names are generated when the techniques go to commercial domains, such as vibration-controlled transient elastography (VCTE)™, Fibroscan™, Fibrotouch™, Virtual Touch™, real-time tissue elastography™ and shear wave elastography (SWE)™. These techniques have all been described in Chapter 6. As different techniques and devices provide different parameters to indicate tissue elasticity with numerous assumptions, it is almost impossible for end-users to understand the fundamentals. When clinicians use an ultrasound scanner to measure blood flow, they clearly know what the value means. However, when the same clinicians obtain a parameter about tissue elasticity, they have to understand much more about tissue biomechanics, which is often beyond what they have been trained. For many end-users, knowledge about tissue elasticity measurement and imaging may come from a salesperson of a specific device, which requires that the salesperson be very knowledgeable and not be biased with the different techniques available.

In 1993, the authors' team began work on the development of novel techniques for measuring tissue elasticity *in vivo*, starting from ultrasound indentation. Since then, the team has developed a series of methods, including ultrasound indentation, water-jet ultrasound indentation, air-jet indentation, vibro-ultrasound shear wave propagation, real-time image-guided transient elastography, ultrasound compression, ultrasound swelling and optical coherence tomography (OCT)-based indentation and suction. They have become one of the many teams, globally, in generating new terms related to tissue elasticity measurement. During substantial collaborations with collaborators working in many different medical and health care fields, they realised the huge gap between the techniques development engineers and health care professionals use. The users tended to simply accept whatever quantitative parameters a device provided and seldom spent time understanding the principle and assumption behind such parameters. In addition, it is very unique that the author's team has been working on techniques to bridge modern ultrasound and optical imaging with conventional elasticity measurement techniques, including indentation, compression and suction tests. This gives an opportunity for them to appreciate even more the huge diversity of tissue elasticity measurement techniques. There are many devices available using indentation, suction, resonant frequency shift, etc., including Myotonometer™, Myoton™, Cutometer™, Ocular Response Analyzer™, Artscan and tissue ultrasound palpation system (TUPS). The principles of these devices are discussed in related chapters in this book together with the constraints and assumptions used.

In addition to the diversity of measurement techniques and devices, tissue elasticity measurement can be further complicated by the complex mechanical behaviours of soft tissues, particularly measured *in vivo*. Soft tissues are actually inhomogeneous, anisotropic, nonlinear, viscoelastic and time dependent (dynamic). Handling complicated behaviours during the measurement of tissue elasticity is a very important topic. If they are not considered carefully, the reliability of measurement results will be questionable. Again, this kind of knowledge is beyond the understanding of many end-users, such as clinicians who want to use elasticity value of tissues for disease diagnosis or tissue assessment. In this book, we briefly discuss the origins of nonlinearity and viscoelasticity of tissues and how to reduce the influence of such behaviours on the measurement. For more comprehensive understanding of the fundamental knowledge about the topic, the readers may refer to Prof. Fung's book.

While the authors focus on a comprehensive review of tissue elasticity measurement *in vivo*, they also give introductions to many techniques available for *in vitro* studies. These *in vitro* methods are not only important for readers to form an overall understanding of tissue elasticity measurement, but also for validating any newly developed measurement techniques. This book covers topics from measurement techniques to clinical applications. Some typical clinical applications are discussed in Chapter 10, but they are by no means exhaustive. The field of tissue elasticity measurement is still rapidly growing, and the future is difficult to predict. The authors propose two future directions for research in this field. One is to standardise the terms and parameters, or even the test protocols used in different fields in the near future. Currently, when describing tissue elasticity, people in different fields and devices based on different techniques may use different parameters, including modulus, Young's modulus, shear modulus, effective modulus, elastic modulus, shear wave speed, stiffness, hardness, firmness, compliance, tenderness, pliability, etc. Each of these parameters can also be defined differently by different research teams. The second is that one technique can be standardised to dominate the field, while devices can be adapted to fit the measuring requirements for different tissues. In this way, the results obtained for the same tissue by different clinicians at different places can be comparable and a standardised protocol can be established.

Acknowledgements

IT WOULD HAVE BEEN IMPOSSIBLE for the authors to complete this book without all the research and development work conducted by the PhD and MPhil students: Minhua Lu, Qing Wang, Congzhi Wang, Clare Chao, Shanica Hon, Shuzhe Wang, Alex Choi, Tak-Man Mak, Sushil Patil, Jiawei Li and Connie Cheng; the authors are grateful for their help.

A group of postdoctoral fellows have also contributed to the research work, including Drs. Tianjie Li, Jiaying Zhang, Match Ko, Lei Tian, Haijun Niu, Carrie Ling, Queeny Yuen, Yongjin Zhou and Jun Shi, together with a group of engineers, including Like Wang, Junfeng He, Zhengming Huang, Louis Li, William Chiu, Jiangang Chen, Guohua Pan and Zhongle Wu. Dr. Tianjie Li also directly contributed to Chapter 9.

Other team members have also made different kinds of contributions to the field, including Dr. Xin Chen, Dr. Jingyi Guo, Dr. Hongbo Xie, Wentao Liu, Jinsheng Fung, Chunhong Ji, Timothy Lee, Dr. Guangquan Zhou, Dr. Weiwei Jiang, Dr. Haris Begovic, Dr. James Cheung, Yi Wang, Kelly Lee, Jinxin Zhao and Yen Law.

The authors also express their sincere thanks to all their collaborators in different fields, and all these collaborations have inspired them to continue improving techniques for tissue elasticity measurement *in vivo*. They are also thankful to Professor Arthur Mak, their mentor in the field of tissue biomechanics, and to Sally Ding, who has contributed her time and efforts in editing the majority of theses and papers generated by the team. Finally, the authors thank the editorial and production staff of CRC Press for their patience, care and cooperation in producing this book and the supports from colleagues of PolyU Technology and Consultancy Company Limited (PTeC) and Interdisciplinary Division of Biomedical Engineering of the Hong Kong Polytechnic University.

Yong-Ping Zheng
Yan-Ping Huang
Hong Kong

Authors

Professor Yong-Ping Zheng earned BSc and MEng degrees in electronics and information engineering from the University of Science and Technology of China, Hefei, Anhui, P.R. China. He earned his PhD in biomedical engineering from the Hong Kong Polytechnic University (PolyU) in 1997. After a postdoctoral fellowship at the University of Windsor, Canada, he joined PolyU as an assistant professor in 2001 and was promoted to associate professor and professor in 2005 and 2008, respectively. He served as the associate director of the Research Institute of Innovative Products & Technologies in PolyU from 2008 to 2010 and has been serving as the head of the Interdisciplinary Division of Biomedical Engineering since its establishment in 2012. Prof. Zheng's main research interests include tissue elasticity measurement and imaging, biomedical ultrasound imaging and wearable sensors for health care. He has published a large number of papers in the field of ultrasound for soft tissue assessment. As chief supervisor, he has supervised 10 PhD students, 6 MPhil graduates and more than 10 postdoctoral fellows. He is on the editorial boards of a number of journals, including *Ultrasound in Medicine and Biology*, *Physiological Measurement* and *Journal of Orthopaedic Translation*. He is a senior member of the IEEE, a fellow of the Hong Kong Institution of Engineers, and secretary-elect of the World Association of Chinese Biomedical Engineers. He holds 7 U.S. and 10 Chinese patents and has another 8 patents pending, many of which have been licensed to industry for commercialisation.

Dr. Yan-Ping Huang is now a teaching fellow at the Interdisciplinary Division of Biomedical Engineering, Faculty of Engineering, the Hong Kong Polytechnic University, Hung Hom, Hong Kong. He earned his bachelor's degree (2002) in electronic engineering and information science from the University of Science and Technology of China (USTC) in Hefei, and MPhil (2005) and PhD (2013) both in biomedical engineering (BME) from the Hong Kong Polytechnic University. After this, he completed one-year postdoctoral training in the Department of Bioengineering, University of Washington (Seattle), before serving in his current position. His main research interests include the measurement and imaging of soft tissue elasticity, using high-resolution medical imaging techniques such as high-frequency ultrasound and optical coherence tomography, in both instrumental development and preclinical investigations, targeting the early diagnosis of diseases of small-scale tissues, including skin, articular cartilage and the microvascular network. He has abundant experience in research and collaboration in the field of BME related to the measurement of soft tissue elasticity *in vivo* and has published quite a number of internationally peer-reviewed papers in the field.

CHAPTER 1

History and Recent Development in Soft Tissue Elasticity Measurement

U SING SOFT TISSUE ELASTICITY for the assessment of different path-ological conditions has been a clinical approach for thousands of years. The most common technique is referred to as 'palpation', i.e. using fingers to feel tissue elasticity (Figure 1.1). The elasticity of soft tissues can be referred to as hardness, stiffness and so on, which represents how much a tissue can be deformed under a certain loading condition. For example, when a tissue is undergoing fibrosis, it becomes stiffer; when there is an oedema, it may become softer. In ancient Greece, palpation was recommended as a method to detect stiffening or pain of the abdomen by Hippocratic physicians (Nicolson 1993). For example, palpation was used as part of the practice for differentiating between ascites and tympanites. Because of the difference in internal fluids, the operator would feel quite differently for the two different pathologies. In the eighteenth century, palpation was also used as a bedside practice for the detection of tumour as recorded in Mogagni's classic work *The Seats and Causes of Diseases as Investigated by Anatomy*: 'and being asked to feel the man's belly, I scarcely perceiv'd any particular tumor elsewhere than in the scrobiculus cordis'.

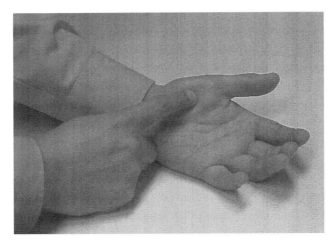

FIGURE 1.1   Illustration of manual palpation of tissue elasticity using a finger. The operator uses the fingertip to sense subjectively and qualitatively the tissue elasticity through the estimation of applied load and induced tissue deformation.

However, measurement of soft tissue elasticity is not an easy task because the elasticity of such a tissue not only depends on the anatomical locations but it also may change with the ageing process. The conventional method using finger palpation is rather subjective and qualitative; therefore it is difficult to use for objective tissue assessment or quantitative treatment outcome measurement. In addition, complex compositions of the soft tissue make it difficult to separate the different elasticity states of different components in the tissue by a finger-tip palpation.

The first device reported for the quantitative assessment of soft tissue elasticity *in vivo* can be traced back to 1912 when Schade developed a mechanical indentation device for the assessment of tissue oedema (Schade 1912). Indentation is a mechanical test method that mimics the operation of hand palpation. However, compared to manual palpation, an indentation test on soft tissue is quantitative rather than qualitative and more objective because it quantifies both the force and deformation for the extraction of quantitative tissue parameters related to tissue elasticity. Since then, indentation measurement has been actively researched using various approaches to measure the force or deformation/displacement separately or simultaneously in order to obtain the mechanical properties of soft tissues. The technical development of indentation was closely related to the advancement of sensor technology. For example, at the very beginning, no electric displacement sensor was available. Then a simple mechanical device was developed for recording the displacement during an indentation test. The working principle of this displacement recording device was very similar to that of a seismograph. The displacement at the indentation site was amplified by a lever, with the other side having a tip pen touching a drum wall. The drum was rotated at a constant speed, and then the time-dependent displacement was recorded continuously on the paper on the drum wall. This kind of displacement recording device was quickly replaced by modern displacement/position sensors such as the linear variable differential transformer (LVDT) or potentiometer (Nyce 2003). Conventional indentation methods, which mainly adopt a force sensor and a displacement sensor for data collection, are described in Chapter 3. Before that, traditional methods for the measurement of tissue mechanical properties *in vitro* are introduced first in Chapter 2. Most of them are standardised mechanical testing methods well used for industrial material testing before they were adopted for biological tissue test. These methods include compression, stretching, shear, inflation, bending, spinning and osmotic swelling. These methods can well be used for specifically prepared tissue samples *in vitro* or *in situ*, but cannot be easily used for *in vivo* because of specific requirements in conditioning the tissue under test.

As the sensors for displacement measurement are usually large in size and indenters require direct contact with tissue, the conventional indentation device is relatively bulky and cannot get information such as the deformation of tissues inside the body. When biomedical imaging methods become available, they can replace the traditional displacement sensor with a better tracing of the dynamic tissue response. Modern novel indentation methods with the incorporation of various advanced biomedical imaging methods are introduced in Chapter 4. More flexibility has been brought with the adoption of modern imaging methods. In addition to indentation, suction measurement is another method suitable for *in vivo* assessment of soft tissue elasticity, though its applications are mainly limited to superficial tissues. Chapter 5 describes the development and applications of suction measurement.

It should be noted that because of the requirement for contact in either force or deformation measurement, in most situations indentation or suction devices can be applied only for the measurement of external soft tissues, such as skin and the whole outside soft tissue layers of the extremities. Elasticity measurement of internal tissue such as liver is still difficult using the indentation or suction method. In this case, indirect measurement techniques such as the shear wave propagation method are more appropriate to measure the internal tissue elasticity. Chapter 6 mainly discusses the indirect elasticity measurement methods. In this situation, a surrogate parameter such as the resonant frequency change in a tactile sensor or shear wave speed in the shear wave propagation method is used to indirectly extract the mechanical properties of the tested tissue.

The conventional tissue elasticity measurement method focuses more on obtaining a quantitative value to represent the tissue stiffness. Since early 1990, the concept of tissue elasticity imaging or elastography has been introduced to map tissue elasticity distribution, which is particularly driven by the diagnosis of cancer, such as breast cancer and prostate cancer. This field has developed rapidly, from initially strain imaging based on quasi-static compression to elasticity mapping based on acoustic radiation force and also shear wave elastography. The techniques to detect tissue disturbance have also evolved from ultrasound imaging initially to magnetic resonance imaging (MRI) and optical imaging. Some elasticity imaging techniques can provide quantitative measurement, but others can only provide semi-quantitative assessment or only the relative distribution of tissue elasticity. This topic related to elasticity measurement versus elasticity imaging is described in Chapter 7.

Though soft tissues have been commonly simplified as homogeneous, isotropic, linear elastic materials, most of them are inhomogeneous, anisotropic, nonlinear, viscoelastic materials in real situations. Soft tissues are structurally arranged in various hierarchies and embedded with plenty of water content. Their mechanical properties can be very different depending on whether they are measured quickly or slowly, *in vivo*, *in situ* or *in vitro*. Related research works are introduced in Chapter 8. When the soft tissues to be measured involve complicated geometry and different substrates, simple analytical solutions for calculating mechanical parameters may not work well or become difficult to achieve. To obtain more accurate intrinsic mechanical properties, finite element analysis and inverse solutions have been extensively studied, which are covered in Chapter 9.

During the last two decades, we have seen great advancements in technique development for the measurement of tissue elasticity, particularly *in vivo*. Some of the newly developed methods have already been commercialised and widely used in clinical setting, such as Fibroscan, a device for the assessment of liver fibrosis based on vibration stimulation and ultrahigh-speed ultrasound measurement. Some techniques have been extensively explored for laboratory and research use. In Chapter 10, some typical applications of tissue elasticity measurement techniques are described, including breast cancer, liver fibrosis, muscle function, brain tissues, eye tissues and others.

While plenty of research and development have been conducted in the area of soft tissue elasticity measurement, particularly for *in vivo* measurement during the last two decades, we observed some trends and dynamic situations in the field. For example, there are many different methods and systems available for the soft tissue elasticity measurement *in vivo*, but the question is which one should be selected for a specific application. Will there be one single technique to be developed that is powerful enough to replace all other techniques and achieve uniform measurement and report in this field? Regarding quantitative results, many different parameters related to biomechanical properties of tissue have been used for describing tissue elasticity, such as strain, stiffness, elasticity, hardness, compliance, firmness, tenderness, Young's modulus, shear modulus, tangent modulus and shear wave speed. Of these parameters, which one do we choose, or is there any limitation or constraint when they are applied to different situations? For researchers, what are the potential directions for research and development in this field? For clinicians, what techniques and parameters should be used for their specific applications? More related questions are briefly discussed in Chapter 11.

CHAPTER 2

Conventional Methods for Soft Tissue Elasticity Measurement *in Vitro*

2.1 INTRODUCTION

The measurement of soft tissue elasticity can be conducted from a 'simple' tissue sample *in vitro*. The tissue sample can be obtained either from a specially prepared animal for laboratory study, or from patients in clinical situations, such as orthopaedic surgery or tissue biopsy for pathological study. The measurement methods for tissue samples *in vitro* include some standard methods that are adopted from materials science, such as compression and indentation tests as well as some special methods such as swelling in articular cartilage tests. The standard testing methods are robust to the test conditions and external factors, and therefore can be used as a gold standard for the elasticity measurement. In materials science, it is very important to measure the elastic properties of metals and plastics. Although soft tissues are quite different from these engineering materials in respect to their functions and behaviours, they are intrinsically similar under some conditions or assumptions, for example when they are treated as simple solids. Therefore, with some adaption, those mechanical testing methods in materials science can be used for testing soft tissues. However, there is abundant water in biological soft tissues, and also some components in the soft tissues are aligned in different directions at different locations, such as the collagen fibres in articular cartilage. The heterogeneity and anisotropy of material properties

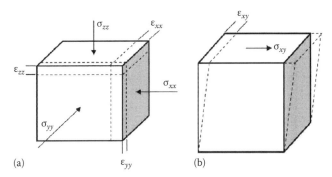

(a)   (b)

FIGURE 2.1 Illustration of strain and stress applied on an infinitesimal block of material. (a) The block is under normal stress from three different directions. (b) The block is under shear stress from one direction along xy plane.

make it very complicated to measure the elastic properties of soft tissue. In practice, it is necessary to simplify the tissue model to get a few but effective parameters of elasticity. Starting from a simple solid model, as shown in Figure 2.1, theory of continuous solid mechanics shows that the stress–strain relation can be expressed as obeying Hooke's law under the infinitesimal strain, and the constitutive equation can be written as (Fung 1965):

$$\varepsilon_{xx} = \frac{1}{E}\left[\sigma_{xx} - \nu\left(\sigma_{yy} + \sigma_{zz}\right)\right] \tag{2.1}$$

$$\varepsilon_{xy} = \frac{1}{\mu}\sigma_{xy} \tag{2.2}$$

where
  x, y and z represent the three orthogonal directions
  ε is the strain
  σ is the stress
  ν is the Poisson's ratio
  μ is the shear modulus
  E is Young's modulus

When the material is assumed to be isotropic and elastic, only two material parameters out of the three are independent, and the relationship can be written as

$$\mu = \frac{E}{2(1+\nu)} \tag{2.3}$$

These fundamental material parameters need to be measured by mechanical tests. These mechanical tests include compression, extension, torsion and bending, which will be introduced subsequently. When soft tissues are assumed to be linear, elastic, isotropic and homogeneous, their elastic properties can be measured by these mechanical testing methods. However, these models are not sufficient for describing the complicated biomechanical properties of soft tissues. More advanced models have been developed to better describe the nonlinear viscoelastic properties of soft tissue. These models are based on the constitutive equations of materials, and related experiment can be conducted to extract the model parameters (Fung 1993a). For example, two commonly used models, namely the hyperelastic model (Miller and Chinzei 1997; Miller-Young et al. 2002) and the quasilinear viscoelastic (QLV) model (Fung 1993a; Ledoux and Blevins 2007; Pai and Ledoux 2011), have been used to predict the nonlinear viscoelastic behaviour of soft tissues in testing. Optimised model parameters can be a better representation of the tissue elasticity.

The advantages of measuring soft tissue elasticity *in vitro* are obvious. First, the experimental conditions, such as the selection of different test methods, can be better controlled so that the measurement may become more reliable and results more accurate. Therefore, it is commonly used as a reference method for validating a newly developed measurement technique. Second, the tissue sample can be prepared in a regular shape so that some commonly used test methods, such as compression and extension, can be directly applied. Third, with an *in vitro* measurement, other tissue characteristics, such as the change of structure, can be obtained through biochemical analysis, and thus the correlation between structural change and mechanical properties can be studied. This point is specifically useful to judge whether the measured mechanical properties are a good reflection of the structural or pathological change of the tested tissue and confirm whether the mechanical properties have the potential for diagnostic purposes.

There are also some obvious disadvantages in the measurement of soft tissue elasticity *in vitro*. Dissected tissues are not live, as in the living organs. The change of its surrounding conditions and the absence of metabolism may make it quite different from its original status in the living body. Therefore, the properties measured *in vitro* may not be a good representation of those of the live condition. For some particular tissues, the change may be quite large, especially for those surrounded by a thin capsule, such as the liver or lymph node. Therefore, for *in vitro* experiments, it is preferable to conduct a test under conditions that are closer to the tissue's *in vivo* status. For some tissues such as the articular cartilage, the tissue properties can be maintained well with cryopreservation. For these tissues, the *in vitro* test can be a common method for characterising the material properties. For different tissues, the situations may be quite different; it is necessary to evaluate the specific situations of the tissue and then decide which is the most appropriate test for the measurement of tissue elasticity *in vitro*. In this chapter, the conventional tests for the measurement of soft tissue elasticity are introduced in the following subsections.

2.2 METHODS FOR MEASUREMENT OF SOFT TISSUE ELASTICITY

In the following sections, methods that are commonly used for the measurement of soft tissue elasticity *in vitro* are introduced. These methods include compression, extension, indentation, suction, inflation, shear, bending, spinning and osmotic swelling. Indentation and suction tests are introduced in separate chapters because these two methods are also common for *in vivo* tests. Therefore, they are not the focus of this chapter. For most of the methods, the tissue elasticity is obtained under the assumption of a simplified model of linear elasticity (homogeneous, isotropic and infinitesimal strain). More complicated models are introduced in Chapter 8.

2.2.1 Compression (Unconfined and Confined)

Compression is the most frequently used method for the measurement of soft tissue elasticity *in vitro*. Although multiaxial compression is possible, it is not very common in soft tissue testing. In an unconfined uniaxial compression test, a regularly shaped tissue sample, for example in a cylindrical shape, is placed between two smooth platens with a larger surface area than that of the sample. The lateral side is allowed to expand freely. The relationship between the load and deformation is then recorded for the calculation of Young's modulus. Assuming that the load is applied along the *x*-axis, which means the stresses in other directions are all zero, Equation 2.1 is simplified to

$$\varepsilon_{xx} = \frac{1}{E}\sigma_{xx} \qquad (2.4)$$

In this case, by assuming a uniform stress–strain relation in the test, Young's modulus of the soft tissue can be obtained as

$$E = \frac{\sigma_{xx}}{\varepsilon_{xx}} = \frac{F/A_0}{\Delta L/L_0} \tag{2.5}$$

where
   F is the compression force
   A_0 is the contact area between the sample and the compressor
   ΔL is the thickness change
   L_0 is the initial thickness of the tissue sample (Figure 2.2)

To prevent buckling and shear during the test, the aspect ratio (diameter/height) of the sample can be chosen to be relatively large. After compression, there will be some expansion of the tissue in the lateral direction. Poisson's ratio ν can also be measured from the test as follows:

$$\nu = -\frac{\varepsilon_{yy}}{\varepsilon_{xx}} \tag{2.6}$$

where ε_{yy} is the strain in the lateral direction, and the sign of minus indicates the difference of change in different directions. The lateral expansion can be measured by a number of methods. Jurvelin et al. (1997) measured this lateral expansion using microscopic observations. Lu et al. (2005) utilised ultrasound to detect the lateral deformation for the measurement of Poisson's ratio. For the compression test, it should be noted that the friction between the platen and tissue sample may affect the test results because this

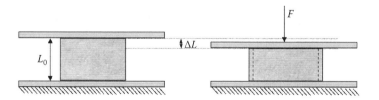

FIGURE 2.2   Illustration of an unconfined compression test. The contact area of the tissue sample is A, and its original thickness is L_0. The tissue is deformed by ΔL under the applied force F.

friction will prevent the lateral expansion of the tissue near the contact surface and cause deviation from the theoretical model of compression (Wu et al. 2004a). The difference increases as the friction coefficient increases. Finite element analysis has shown that when the friction coefficient is 0.5, the difference of stress at certain strain can be as large as 50% and the effect is not negligible (Wu et al. 2004). In real practice, polished platens with small friction (Miller-Young et al. 2002) or platens covered with polytetrafluoroethylene (PTFE) or lubricants (Lu et al. 2005; Rashid et al. 2012) can be used to reduce the effect of friction during the measurement.

Confined compression is another test that has been commonly used for the measurement of soft tissue elasticity. It is used because it can better describe the situation of soft tissue in a real situation *in vivo*, where it is subjected to resistance from its surrounding tissues, unlike free of resistance in the case of unconfined compression. When its surroundings are confined (Figure 2.3), the lateral strain is zero, and then according to Hooke's law in another form

$$\sigma_{xx} = \frac{(1-v)E}{(1+v)(1-2v)}\varepsilon_{xx} + \frac{vE}{(1+v)(1-2v)}\left(\varepsilon_{yy} + \varepsilon_{yy}\right) = H_A\varepsilon_{xx} \quad (2.7)$$

where $H_A = \dfrac{1-v}{(1+v)(1-2v)}E$ is called the 'aggregate modulus', representing the elastic properties of an ideal elastic material when confined laterally. When the tested material is completely noncompressible, that is, $v = 0.5$, the aggregate modulus will approach infinity, which hardly happens in real situation; when the material is completely compressible, that is, $v = 0$, the aggregate modulus will be equal to Young's modulus. Therefore, the aggregate modulus is a direct indication of the material compressibility.

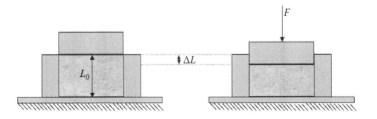

FIGURE 2.3 Illustration of a confined compression test. The contact area of the tissue sample is *A*, and its original thickness is L_0. The tissue is deformed by ΔL under the applied force of *F*.

Confined compression is used to test mainly articular cartilage (Mow et al. 1980; Ateshian et al. 1997; Schinagl et al. 1997; Soltz and Ateshian 1998; Chen et al. 2001; Korhonen et al. 2002; Knecht et al. 2006) and also some other soft tissues including the meniscus (Joshi et al. 1995), intervertebral disks (Yao et al. 2002) and skin (Wu et al. 2003). A biphasic model can be assumed for the tested tissue, and different materials can be selected for the compressor and surrounding walls for the control of the fluid flow during the test (Mow et al. 1980). For example, in a confined test of articular cartilage, impermeable material is chosen for both the wall and the bottom, but a porous compressor is used on the top of the tissue sample, allowing the fluid to freely move out of the tissue during compression. Associated with a creep or force relaxation test, the permeability of the tissue can also be measured from confined compression in addition to the aggregate modulus and Poisson's ratio (Mow et al. 1980).

Tissue deformations under compression are conventionally obtained through the displacement measurement of the compressor. Alternative methods for deformation measurement have also been reported in the literature, considering that the conventional method cannot resolve deformations of different layers of tissue under compression. Ultrasound imaging has been used for the internal strain measurement of the articular cartilage under axial compression (Zheng et al. 2004) and applied for the tracking of transient (Zheng et al. 2005) and equilibrium strains (Zheng et al. 2002). The usability for characterisation of depth-dependent elasticity has been revealed also for degenerated tissue (Zheng et al. 2001; Qin et al. 2002). An ultrasound study has confirmed that tissues inside the cartilage undergo a strain relaxation phenomenon during a prolonged compression (Zheng et al. 2005). It has also been demonstrated *in situ* that a soft superficial tissue layer, generated by depleting proteoglycans (PGs) with trypsin treatment, can be assessed with ultrasound measurement (Zheng et al. 2001). Fortin et al. (2003) used ultrasound to detect transient lateral deformation of cartilage tissues under an axial compression with two platens and demonstrated time-dependent Poisson's ratios of the cartilage. Such evaluations cannot be achieved using the conventional strain measurement under a compression test.

2.2.2 Extension

The test direction in an extension (elongation) test is opposite that of the compression test. Extension is also commonly used for the measurement of some soft tissues, including blood vessels (Steiger et al. 1989; Ozolanta et al.

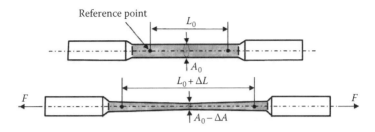

FIGURE 2.4   Illustration of elongation (extension) test for a sample with original length L_0 and original cross-sectional area A_0.

1998; Monson et al. 2003; Holzapfel 2006), ligaments (Yoganandan et al. 2000; Gupte et al. 2002), tendons (Almeida-Silveira et al. 2000; Ng et al. 2003; Revel et al. 2003) and skin (Pan et al. 1998; Eshel and Lanir 2001), because these tissues undergo some kind of extension when taking part in the normal activities of the body. Compared to the compression test, for the elongation test a relatively long strip of soft tissue is prepared and clamped by grips at the two ends (Figure 2.4). Sandpaper can be placed on the surface of the grips to increase the friction to prevent the relative motion of the sample and grips. Although there will be some inhomogeneity in the strain distribution of the tissue sample during extension, the stress–strain relationship can be approximated by the average stress and strain values at the centre of the specimen, based on the Saint-Venant principle. Therefore, the equation to calculate Young's modulus is similar to that of the compression test. The cross-sectional area of the sample needs to be measured to calculate the stress applied in the tissue. Extension along different directions can be applied to measure the anisotropy of the tissue elasticity, as in the case of blood vessels: one along the vessel direction and the other in the radial direction (Ozolanta et al. 1998; Monson et al. 2003). The extension test is especially appropriate for the testing of collagen fibrils in connective soft tissues (Gentleman et al. 2003).

2.2.3 Inflation

The inflation test is developed to measure the elasticity of a tubular or spherical soft tissue where it is internally inflated by a fluid, such as a blood vessel (Kim and Baek 2011) or cornea (Elsheikh et al. 2007). For this test, the soft tissue needs to be enclosed in a pressurised chamber, and the mechanical behaviour of the tissue is detected using a medical imaging system or cameras to extract its mechanical properties (Figure 2.5). A shell theory can be generally adopted for analyzing the data; for example, in the case

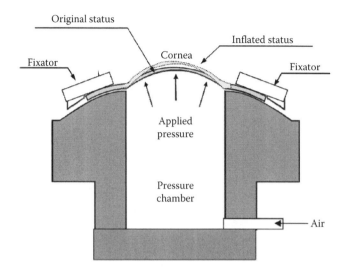

FIGURE 2.5   Illustration of the inflation test of a cornea sample.

of the inflation test for a cornea, Young's modulus in the circumferential direction can be calculated as (Elsheikh et al. 2007)

$$E = \frac{pR^2}{2t\Delta H}(1-v)\left(1-e^{-\beta\eta}\cos(\beta\eta)\right) \tag{2.8}$$

where
   p is the applied pressure
   R is the radius of the corneal meridian surface
   t is the average corneal thickness
   ΔH is the apical rise after inflation
   v is Poisson's ratio
   η is half of the corneal surface angle
   $\beta = \sqrt{R/t} \cdot \sqrt[4]{1-v^2}$ is a constant related to the geometry

Inflation can also be combined with the extension test to mimic the physiological loading situation of the blood vessel during circulation (Kim and Baek 2011). The circumferential and longitudinal elasticity of the octopus aorta was studied using the inflation test (Shadwick and Gosline 1985).

2.2.4 Shear

Biomechanical soft tissues are associated with not only axial deformation but also shear deformation in their daily activities. For example,

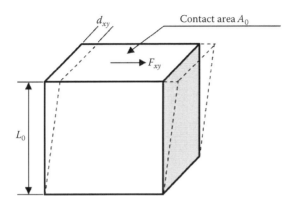

FIGURE 2.6   Illustration of the pure shear test using a shear force applied on the upper surface to cause a lateral displacement.

the different layers of the myocardium undergo shear stress, which is demonstrated to be one of the mechanisms for systolic myocardial wall thickening (LeGrice et al. 1995). Shear is also one of the main reasons for traumatic injury caused to the brain in an accident (Margulies et al. 1990). Therefore, it is also important to measure the shear elastic properties of the soft tissues. Two types of shear tests are normally used. The first is the pure shear test, while the second is the torsion test. In the former (Figure 2.6), a shear force is applied to the upper surface of the tissue sample, and then the lateral displacement (d_{xy}) is measured as the shear deformation with respect to its initial height (L_0). Assuming the shear force is F_{xy} and the contact area is A_0, the shear modulus of the tissue can be calculated as

$$\mu = \frac{\sigma_{xy}}{\varepsilon_{xy}} = \frac{F_{xy}/A_0}{d_{xy}/L_0} \tag{2.9}$$

Some shear testing devices have been proposed in the literature for the measurement of shear properties. Dokos et al. developed a tri-axial measurement shear test device for the measurement of soft tissues (Dokos et al. 2000), and later they used this device to measure the anisotropic elastic properties of the passive ventricular myocardium (Dokos et al. 2002). Tanaka et al. also utilised the pure shear test to measure the mechanical behaviour of the temporomandibular joint disk in different directions and frequencies (Tanaka et al. 2003). Setton et al. (1995) showed that the mechanical behaviour of the articular cartilage in shear was altered by transection of the anterior cruciate ligament.

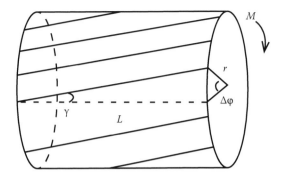

FIGURE 2.7   Illustration of torsion test.

The second method for a shear test is the torsion test (Figure 2.7). Normally, a cylindrical tissue sample is prepared and a torque is applied to the two ends of the sample to rotate the tissue through a certain angle. Assuming the twist angle to be $\Delta\varphi$ for a sample with height L and radius r, the maximum shear at the circular ridge is

$$\varepsilon_{xy} \approx \gamma = \frac{r\Delta\varphi}{L} \tag{2.10}$$

where γ is the angle of shear. The related stress at this point is

$$\sigma_{xy} = \frac{Mr}{J} \tag{2.11}$$

where
   M is the torque applied to the sample
   J is the polar moment of inertia for the tested samples (Beer et al. 2006a)

Finally, the shear modulus of the tested sample can be calculated as

$$\mu = \frac{\sigma_{xy}}{\varepsilon_{xy}} = \frac{ML}{J\Delta\varphi} \tag{2.12}$$

For simplification, sometimes the torque/twist angle ratio (M/φ) is directly used as an indicator of the tissue stiffness (Zdero et al. 2009). The standard torsion test is appropriate only for some relatively stiff tissues such as the articular cartilage (Setton et al. 1995) and ligament (Ruland et al. 2008;

Zdero et al. 2009), as it is not so easy to keep the shape of the tissue under torsion for very soft tissues.

2.2.5 Bending

Bending can also be used to measure the elasticity of soft tissues (Nicosia 2007), including the arterial wall (Yu et al. 1993; Xie et al. 1995) and the aortic valve leaflet (Mirnajafi et al. 2005, 2006), because bending is a dominant mode of deformation experienced by heart valves in their normal physiological functions. Young's modulus of the specimen can be obtained through a typical three-point bending test (Figure 2.8) (Beer et al. 2006b):

$$E = \frac{Mr}{I} \tag{2.13}$$

where

   M is the bending moment applied to the sample
   r is the radius of curvature after deformation
   I is the moment of inertia for the tissue sample; for a typical cuboid sample with a cross-sectional square of width b, the moment of inertia is $I = b^4/12$

Yu et al. (1993) studied the neutral axis of the blood vessel in bending and found that it existed at one-third of the wall thickness. Furthermore, they found that the intima media layer was one order larger in stiffness than the adventitia (Yu et al. 1993). Roy et al. (2004) extended the bending test to characterise the large deformation behaviour of native and tissue-engineered cartilage. Through the study, they found that the stiffness of the auricular cartilage was smaller than that of costal cartilage and that tissue-engineered cartilage was softer than native cartilage.

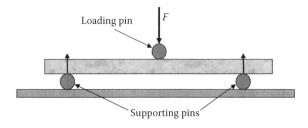

FIGURE 2.8   Illustration of the three-point bending test of a sample.

2.2.6 Spinning

The spinning test was proposed specifically for the measurement of lens stiffness (Fisher 1971). The deformability of the lens in accommodation is significantly affected by its stiffness, and the stiffening of the lens is hypothesised to be the most probable reason for poor near-sight vision in elderly people, called presbyopia. Therefore, it is important to characterise the mechanical properties of the lens. The spinning test is used to mimic the stress endured by the lens during an accommodation process. During the test, the lens is placed at the centre of a spinning support (Figure 2.9). The lens is assumed to be of an elliptical shape, and the change of the long and short axes can be tracked by a camera or microscope with fine spatial resolution. Then the tissue stiffness at the equatorial and polar sides can be calculated as

$$E_E = \frac{a^3 \rho \omega^2}{8 \delta_a} \tag{2.14}$$

$$E_P = \frac{7 a^2 b \rho \omega^2}{24 \delta_b} \tag{2.15}$$

where
   a and b are the equatorial radius and the distance of the anterior pole to the equatorial plane, respectively, of the accommodated lens
   ρ is the mass density
   ω is the angular rotation speed (in radians)
   δ_a and δ_b are the change of a and b of lens profile due to high-speed rotation

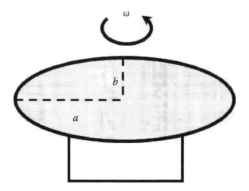

FIGURE 2.9   Illustration of spinning test for a lens.

Later, Burd et al. (2011) proposed an improved spinning test and a finite element analysis for a better measurement of the lens elasticity (Burd et al. 2011). Recent studies using this spinning test clearly showed the change of lens elasticity with age and the existence of elasticity difference between nucleus and the cortex (Chai et al. 2012; Wilde et al. 2012).

2.2.7 Osmotic Swelling

In tissue physiology, it is well known that a Donnan osmotic pressure exists when there is a difference of ion concentrations across a semipermeable membrane. The osmotic swelling pressure has been demonstrated to be one of the main reasons for counterbalancing external forces applied on the soft tissues (Lu et al. 2006). The advantage of this method is that no external force is necessary to be applied to the tissue, and the deformation is induced internally using the osmotic swelling pressure. This method has been applied broadly in the study of the articular cartilage (Basser et al. 1998; Setton et al. 1998; Narmoneva et al. 1999, 2001, 2002; Bank et al. 2000; Flahiff et al. 2002, 2004; Hattori et al. 2009). When the concentration of the immersing saline solution is changed, a new osmotic pressure will be produced and then new balance can be achieved by the extension of collagen fibres. The osmotic swelling test can be used to measure the stiffness of the solid material of the cartilage. For calculation, the stress is obtained from the Donnan pressure, which needs information about the fixed charge density, water concentration and saline concentration at different layers of the cartilage. For strain, it can be quantified using a high-resolution camera (Narmoneva et al. 2001), but in this method the cartilage need to be cut from the testing site and then the strain can only be detected from the side because of limitations of direct observation through a camera.

Tepic et al. utilised ultrasound to observe the deformation of the cartilage under osmotic pressure (Tepic et al. 1983). Only the reflection from the cartilage surface was used to detect the swelling, and no observation was performed for the internal cartilage. High-frequency ultrasound is very useful to detect the motion of internal structures using a motion-tracking method, and therefore it is very useful for extracting the mechanical properties of the cartilage at different layers. Wang et al. utilised a 50 MHz ultrasound transducer to conduct biomicroscopy of the articular cartilage and successfully detected the different situations of deformation at different layers in the swelling test (Figures 2.10 and 2.11). Based on a four-parameter triphasic model, the distribution of aggregate modulus at different depths of the cartilage was obtained (Wang 2007; Wang et al. 2007). The results confirmed

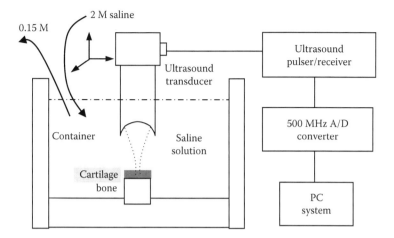

FIGURE 2.10 Illustration of an ultrasound biomicroscopy system for monitoring osmotic swelling of articular cartilage samples. The sample is originally submerged in 0.15 M saline and is then changed to 2 M saline, which induces swelling as observed in Figure 2.11.

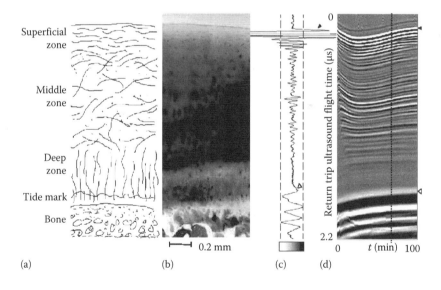

FIGURE 2.11 Typical osmotic swelling phenomenon of articular cartilage observed by ultrasound biomicroscopy imaging. (a) Illustration of the depth-dependent collagen fibre orientation in articular cartilage. (b) Histological image showing the depth-dependent proteoglycan distribution in articular cartilage. (c) A-mode ultrasound signal indicating different cartilage tissue interfaces. (d) M-mode ultrasound imaging showing the dynamic strains of articular cartilage at different depths. (From Wang, Q. et al., *J. Biomech. Eng. Trans. ASME*, 129, 3, 2007.)

the earlier finding that the aggregate modulus increased with increased depth in the cartilage (Schinagl et al. 1997).

2.3 SUMMARY

Because of the difficulty in measuring elasticity of soft tissues *in vivo*, many specific techniques for testing the mechanical properties of such tissues *in vitro* have been developed. These tests can be performed under relatively simple boundary conditions, and then constitutive mechanical parameters can be easily obtained. Therefore, some tests have become the de facto gold standards for the measurement of soft tissue elasticity, which are also, due to their simplicity, a suitable starting point for studying the elasticity of soft tissues from scratch. The elasticity parameters obtained can be used (1) to study the fundamental biomechanics of the soft tissue for modeling, (2) as input into a computer for accurate simulation of the soft tissue in natural physiology or in response to external loading, (3) to conduct a feasibility study to verify whether it is possible for diagnosis, and if confirmed, then such a study is very useful for future development of methods of *in vivo* measurement and (4) to evaluate the feasibility of assessing the treatment effect using elasticity change pre- and post-treatment. These measurement techniques have broad applications in various studies of soft tissue, and they are indispensable, both at present and in future, for research on soft tissue biomechanics, diagnosis and treatment.

CHAPTER 3

Indentation Measurement of Soft Tissue Elasticity

3.1 INTRODUCTION

Hand palpation is commonly seen in clinical practice to assess the pathological conditions of soft tissues, such as in diagnosis of breast and prostate tumours as well as tissue oedema and fibrosis. One of the main objectives of hand palpation is to assess the firmness or tenderness of the tissues and then judge whether there is abnormality associated with the tissue. The operator uses his/her finger to compress the tissue in a self-controlled pace and feels how the tissue responds to deformation. Qualitative or semi-quantitative scores for the tissue firmness can be given based on the palpation feeling of the operator. However, this method suffers from the subjectivity of qualitative assessment, and the results rely heavily on the experience of the operator. Therefore, indentation, as a variant of hand palpation, has been adopted as an objective measurement method in the field for the measurement of soft tissue elasticity.

Indentation, when used as a mechanical testing method, can be defined as a general method to measure the mechanical properties of materials by inducing a notch or recess in the tested object and then analysing the force–deformation relationship. The size of the indenter is smaller than the compressed object, which is the most important difference between indentation and compression. In traditional mechanical engineering, indentation has been widely used to measure the stiffness of various materials including metals, plastic, rubber, foam and elastomers. Some international standards have been defined for testing hardness and other material properties of metallic materials using indentation methods, such as ISO14577-1: Metallic materials—Instrumented indentation test for hardness and materials parameters—Part I: Test method. Commonly used stiffness parameters include the Vickers hardness, Rockwell hardness and the Brinell hardness. In recent years, the instrument-based indentation techniques, including nanoindentation, have been developed (Oliver and Pharr 1992; VanLandingham 2003; Fischer-Cripps 2006). Because the requirement for preparing a specimen for an indentation test is simpler than that for a traditional standard test such as compression, it is useful not only *in vitro* but also *in vivo* using a noninvasive test, and this test has been broadly applied for the measurement of soft tissue elasticity. The earliest application of indentation for tissue assessment, which was used for the test of skin elasticity, was proposed in the 1910s (Schade 1912). At that time, the indentation system was bulky with manual operation, and the method for recording data was also very primitive, which made its applications very limited. However, with the fast development of modern techniques in electronics, computers, sensing, automatic control and signal processing, the indentation techniques for biological soft tissues also have developed rapidly. Until now, quite a number of indentation systems have been developed that are applicable in various testing conditions. In the following section, the development of indentation techniques used for soft tissue elasticity measurement is reviewed.

In 1912, Schade reported an indentation system that was used to measure the elasticity of skin and underlying soft tissues (Figure 3.1). This might be the first report in this field that was related to the application of indentation test for characterising soft tissue elasticity. Although primitive, this system was designed in an artful way. The system used a hemispherical indenter to dent the tissue. The indenter was fixed to one side of a lever, and the displacement of the indenter was amplified at the other side of the lever. The amplified displacement of the tissue surface induced by the indenter was then recorded by a pen with ink on a drum that rotated with time. The indentation force was applied to the soft tissue by a fixed weight placed on top of the indenter. In order to minimise the effect of tissue motion induced by the subject during the test, an additional support was placed on the surface of the unaffected tissue to record the movement of the intact tissue, which was subtracted from the recorded deformation.

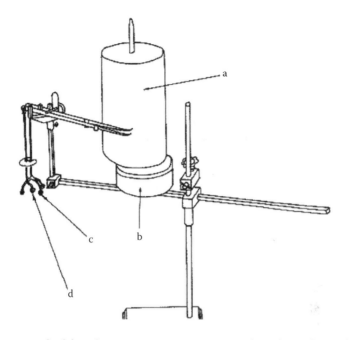

FIGURE 3.1   Schade's indentation apparatus. a: Recording drum, b: mechanical rotating device, c: reference tips, d: indenter. (From Schade, H., *Z. für die Exp. Pathol. Ther.*, 11, 369, 1912.)

The phenomenon of creep after the loading of the weight and the recovery after unloading was observed during the experiment. Results showed that oedematous tissue reached an equilibrium status after adapting to the loading or their recovery in a much longer time, which was understandable on the basis of normal symptoms exhibited in oedema. Kirk et al. improved this method about 40 years later using smoother material on the data recorder to reduce the friction between the pen tip and the drum (Kirk and Kvorning 1949; Chieffi 1950; Kirk and Chieffi 1962). The improved system was applied to study the change of elasticity with ageing and the effect of steroid treatment on skin elasticity. Lewis et al. (1965) then used a skin caliper to study skin oedema. A typical skin caliper is shown in Figure 3.2. Assuming a fixed force was applied, the caliper clamped a fold of the skin and recorded the change of tissue thickness with time. Because of the existence of water, there was a progressive creep of the skin thickness during the caliper indentation. The time to the final balance could be used as a characteristic parameter to compare normal and oedematous tissues. The indentation test conducted by the skin caliper was not appropriate to study the elasticity of the underlying muscle layers, and the

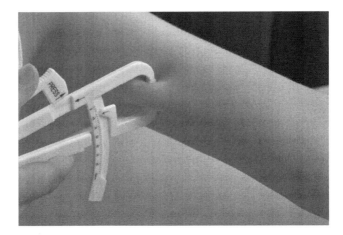

FIGURE 3.2   Typical skin caliper for measuring the skin/fat layer of an arm.

results were also affected by quite a number of factors including the initial thickness and shape of the tissue fold. The reliability of measurement was difficult to guarantee, and therefore it could not be used as a standard tool for the measurement of skin elasticity.

The disadvantages of Schade's indentation system included the following: The indentation force could not be adjusted and measured continuously, a complicated lever system was used for the amplification of the displacement of the pen tip and a physical drum was used to record the tissue deformation. With fast developments in electronics and industrial manufacturing techniques, various new types of sensors have been emerging for recording the physical parameters including force and deformation. Along with better control techniques by the use of personal computers, the measurement reliability and accuracy have been greatly improved for the indentation technique. Ziegert and Lewis (1978) adopted a force sensor and a linear variable differential transformer (LVDT) to measure the indentation force and deformation continuously during a test. Through the study, reliability of the measurement, locational variance and the difference between quasi-static and dynamic test of the soft tissue elasticity were evaluated. Bader and Bowker (1983) used a cylindrical indenter to test the elasticity of skin and subcutaneous soft tissue, in which a potentiometer was utilised to measure the deformation after step force loading and unloading. Test on soft tissue *in vivo* showed that the mechanical properties varied with the test site, sex and age of the subjects, and a simple viscoelastic model also showed that the viscosity of the soft tissue varied with age. Since then, the traditional indentation has been used with quantitative analysis to study the intrinsic mechanical properties of soft tissues. Horikawa et al. (1993) reported an indentation system for the measurement of muscle elasticity that utilised laser reflection to measure the deformation. Special attention was paid to laser safety in this study, and it was found that the stiffness of deltoid muscle increased significantly with the increase of the arm abduction angle in a standing position. Mak et al. (1994) developed a motor-driven indentation system to measure the elasticity of soft tissues of the lower limb. The elasticity of soft tissues between normal and amputated subjects was compared. Pathak et al. (1998) further developed a more portable device that could be mounted in prosthesis for the measurement of elasticity of soft tissues of a limb using rate-controlled indentation. Rome and Webb (2000) designed a special indentation device for the heel, and then compared the stiffness of the heel between subjects with and without heel pain (Rome et al. 2001). Klaesner et al. (2001) developed an indentation system to measure the stiffness of the plantar soft tissues. Table 3.1 lists the indentation systems that have been used for the measurement of the mechanical properties of soft tissue in the literature. Advantages and disadvantages of each system are also given.

The durometer, which has been widely used in materials science, is also adopted in this field for the measurement of soft tissue elasticity (Figure 3.3). Peck and Glick (1956) utilised a durometer and a tonometer to measure the hardness of keratin *in vitro* and *in vivo*, respectively. They studied the effect of hydration on the flexibility of keratin and found that dehydration caused a significant hardening of the keratin. The mechanism of the durometer or tonometer is to apply a fixed force on the knob and then observe the reading of the scale, which indicates the deformation of the tissue. The reading on the scale represents the relative stiffness of the tissue, with a larger value indicating a larger stiffness. This type of durometer is portable, and therefore is convenient, especially for the study of skin elasticity. Examples included the evaluation of skin elasticity or change after therapy under the pathology of various skin diseases or after burns. Hillerton et al. (1982) utilised a Vickers indenter to measure the stiffness of insect cuticle. Katz et al. (1985) utilised the tonometer to evaluate the effect of a pressure garment on the stiffness of a hypertrophic scar. Results showed that the skin stiffness decreases after the treatment with a pressure garment. Falanga and Bucalo (1993) measured the skin stiffness using a durometer on patients with scleroderma. The relationship between the stiffness measured by

TABLE 3.1 Summary of Reported Indentation Systems in the Literature

References	Main Characteristics of the Indentation Systems	Tested Soft Tissues	Advantages (A) and Disadvantages (D)
Schade (1912), Kirk and Kvorning (1949), Chieffi (1950) and Kirk and Chieffi (1962)	Hemispherical indenter with a fixed load; deformation amplified by a lever and recorded with pen ink	Medial surface of tibia	(A) The earliest indentation system for study of soft tissue elasticity (D) Very primitive system with fixed indentation load and old data recording method
Lewis et al. (1965)	Utilised a skin caliper to observe the thickness change under a fixed clamping force. A squeezed part of soft tissue was clamped by the caliper and the tissue was not tested in a natural status	Skins at various body sites including upper and lower extremities, hand and foot	(A) A caliper is widely available for test (D) Results were affected by quite a number of factors (D) Mainly includes the study of skin layer, but not deep layers such as muscle
Kydd et al. (1974)	The initial thickness and deformation of soft tissues were measured by an ultrasound transducer; force measured by strain gauge	Antero-medial surface of tibia	(A) Deformation was measured by ultrasound signals (D) No synchronisation for force and deformation
Ziegert and Lewis (1978)	The indentation system was installed on a frame rigid to the tissue. Utilised LVDT to measure the displacement and strain gauge to measure the loading	Soft tissue covering the antero-medial tibia	(A) Standard parts LVDT and force sensor were used for measurement of force and deformation (D) The system was designed for indentation in the vertical direction

(Continued)

TABLE 3.1(*Continued*) Summary of Reported Indentation Systems in the Literature

References	Main Characteristics of the Indentation Systems	Tested Soft Tissues	Advantages (A) and Disadvantages (D)
Bader and Bowker (1983)	Utilised a potentiometer to measure the displacement	Anterior soft tissue of forearm and thigh	(A) Different models were used for analysis (D) Loading cannot be adjusted continuously
Fischer (1987a,b)	A tissue compliance meter (TCM) was developed. The deformation was measured by the sliding of a disk on the indenter shaft; load measured by a force gauge	Muscles on the back	(A) Manual operation at ease (D) Can be applied only on large area with the big ring of 8 cm in diameter
Mak et al. (1994)	A computer-controlled motor was used to achieve automatic indentation	Soft tissues below and near the knee joint	(A) Automatic indentation (D) System in a large profile, only for vertical indentation
Ferguson-Pell et al. (1994)	The indentation force could be controlled by a pneumatic valve using a force feedback system and corresponding deformation observed	Skin	(A) Easy to achieve the creep function (D) No convenient method for deformation measurement
Lyyra et al. (1995)	Arthroscopic indentation system with a fixed indentation depth; indentation force used as a stiffness index	Articular cartilage	(A) Able to be used in arthroscopic operation (D) No continuous measurement of force and deformation; unable to measure cartilage thickness
Vannah and Childress (1996)	Plane-ended cylindrical indenter with load cell and LVDT to measure the indentation force and deformation	Muscular tissues at the calf region	(A) Portable for testing (D) No measurement of initial tissue thickness

(Continued)

TABLE 3.1 (*Continued*)   Summary of Reported Indentation Systems in the Literature

References	Main Characteristics of the Indentation Systems	Tested Soft Tissues	Advantages (A) and Disadvantages (D)
Pathak et al. (1998) and Silver-Thorn (1999)	Used a linear actuator to drive the indenter; operation could be rate controlled	Soft tissues of lower limbs	(A) Indentation rate easily adjustable (A) Incorporated into prosthesis for operation (D) No measurement for thickness
Vannah et al. (1999)	Used an air pump to achieve cyclic indentation, and the deformation was sensed by an electromagnetic sensor	Soft tissue of lower leg	(A) Able to achieve stiffness mapping (D) No intrinsic parameter obtained
Rome and Webb (2000) and Rome et al. (2001)	An indentation system specifically designed for foot tissue measurement	Plantar soft tissues	(A) Specialised system for foot measurement (D) No measurement of initial thickness
Klaesner et al. (2001, 2002)	Utilised a 3D spatial localisation system to measure the indenter displacement	Plantar soft tissues	(A) Multiple freedom in operation (D) Not quite portable (D) No measurement of initial thickness
Lu et al. (2009c)	A pen-sized indenter with an electromagnetic spatial sensor at the tip	Abdomen soft tissues	(A) System very portable (A) Capable of measuring soft tissue with large thickness (D) Spatial signal easily affected by metal components

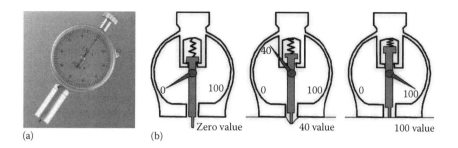

FIGURE 3.3 (a) Typical durometer. (b) Illustration of the working principle of the durometer.

the durometer and that of clinical score was analysed, and it was found that the increase of the disease score was associated with a significant increase of skin stiffness. These devices could also be used for studying oedema (Chen et al. 1988; Mirnajafi et al. 2004); their reliability and effectiveness in studying burn scar or skin diseases have been evaluated (Corica et al. 2006; Lye et al. 2006; Merkel et al. 2008). Recently, an electronic tonometer was reported for the assessment of skin sclerosis (Dugar et al. 2009). The tonometer has been proposed as a clinical tool to evaluate scar flexibility by some researchers, but attention should be paid when applied to soft tissue covering bony prominences (Verhaegen et al. 2011). In addition, a number of tonometer devices have been reported for the assessment of muscle elasticity, commonly called a 'myotonometer', which has a larger indentation head in comparison with those used for skin. It should be noted that the durometer or tonometer uses a stiffness value without unit or the deformation value directly as the measurement results, which are not intrinsic material properties of the measured soft tissues and depend on many factors, including tissue thickness and geometry, indenter size, underlying bony structure, etc. Therefore, the generalisation of the measured results is questionable when measurements are from different centres or different devices need to be compared. When indentation devices are available and data are collected from the measurement, the next important step is how to analyse these data to obtain intrinsic mechanical properties based on a feasible test model, which is described in the following section.

There is a specific category of tonometers designed for measuring the intraocular pressure (IOP) of the eye (Figure 3.4). Starting from an earlier mechanical tonometer, named as 'Goldmann Applanation Tonometer' (Goldmann and Schmidt 1957), there are now different types of electronic

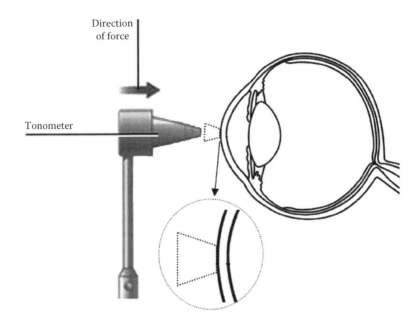

FIGURE 3.4   Illustration of measuring intraocular pressure (IOP) using a tonometer. When the cornea surface is flattened, the pressure applied from the tonometer is assumed to be equal to the IOP.

tonometers available in the market. Its basic principle is that the pressure applied on the compressor equals the IOP when the cornea surface is flattened. However, the accuracy of the tonometer measurement of IOP is affected by inter-individual variations of corneal properties, including corneal thickness, curvature and elasticity (Liu and Roberts 2005). Recently, there has been increased research interest in the measurement of corneal elasticity, and a number of different techniques have been developed, which will be introduced in the subsequent sections. One of these was directly inspired by the tonometer, and it used a small indenter with a diameter of 2 mm to indent corneal tissue to obtain its force–deformation relationship, which was then used to extract Young's modulus (Ko et al. 2013, 2014). A model was derived to calculate Young's modulus from the indentation force and cornea deformation, which is introduced in Section 3.2.4.

3.2 ANALYSIS METHODS FOR TRADITIONAL INDENTATION

After an indentation test, normally a force–deformation curve can be obtained to represent the mechanical behaviour of the soft tissue. However, to calculate the stiffness parameters from the indentation test with the various associated boundary conditions is not an easy task. The simplest way to tackle this problem is to treat the soft tissue as a type of single-phase solid material. When the material is further assumed to be isotropic, homogeneous and linear elastic, some analytical formulae can be used to extract the material properties. This type of single-phase model is useful to represent the elastic properties of the tissue at the instantaneous or equilibrium state of the test. The traditional palpation score used in clinical diagnosis is the reflection of such an average elasticity. However, it is known that there is abundant water in biological soft tissue, which is normally larger than 50% of the body weight. Therefore, a biphasic model is naturally feasible to be proposed for describing the mechanical behaviour of soft tissue. The biphasic model has been widely used in the analysis of an indentation test on articular cartilage. In certain circumstances, an additional phase such as the ion phase in the articular cartilage has been added to the biphasic model, and therefore a triphasic model becomes possible. In addition to the analysis from a specific multiphasic material model, some nonlinear viscoelastic models have also been proposed to describe the relationship between force and deformation in the indentation test. These models consider not only the material viscosity but also the nonlinearity of elasticity induced by a large deformation in the test. In practice, sometimes, the mechanical behaviour of the tested tissue becomes too complicated because of both the complicated material properties from the soft tissues themselves and the complex boundary conditions, so that it is not possible for an accurate theoretical analysis. Finite element analysis (FEA) utilises the large-scale and fast computational capability of computers to simulate the change of soft tissues under an external force and is quite useful for indentation analysis. FEA for the analysis of the indentation test is briefly mentioned in later sections of this chapter, and more details of FEA for the measurement of soft tissue elasticity are presented in Chapter 9.

3.2.1 Single-Phase Analysis of Indentation Test

When the soft tissue is treated as an isotropic, homogeneous, linear and elastic material, the indentation test belongs to the contact problem as proposed and solved in classic mechanics. The soft tissue is assumed as a semi-infinite elastic material bonded to an incompressible bottom plate, and its thickness is much larger than the diameter of a plane-ended cylindrical indenter ($h \gg a$, Figure 3.5). When it is assumed that no friction exists between the indenter and the soft tissue, according to the Hertz contact theory (Timoshenko and Goodier 1951; Johnson 1985),

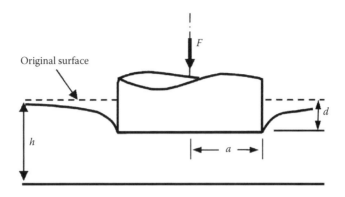

FIGURE 3.5 Soft tissue layer indented by a cylindrical indenter.

the relationship between Young's modulus and force/deformation is as follows (Waters 1965):

$$E = \frac{1-v^2}{2a} \cdot \frac{F}{d} \tag{3.1}$$

where
    E is Young's modulus
    v is Poisson's ratio
    a is the radius of the cylindrical indenter
    F is the indentation force
    d is the deformation

If constant values for Poisson's ratio and the indenter radius are used, Young's modulus is linearly proportional to force/deformation and, therefore, the force/deformation ratio can be directly used to indicate the stiffness of soft tissues in some specific applications.

3.2.1.1 Effect of Thickness and Poisson's Ratio
When the size of the indenter is comparable to the initial thickness of the tissue, then Equation 3.1 should be modified to obtain an accurate evaluation (Waters 1965; Hayes et al. 1972). Hayes et al. (1972) solved the problem by adding a scaling factor into the equation (Figure 3.5):

$$E = \frac{1-v^2}{2a\kappa\left(v, a/h\right)} \cdot \frac{F}{d} \tag{3.2}$$

where κ is a scaling factor related to Poisson's ratio (ν) and indenter radius/tissue thickness (a/h), the latter of which is called as the 'aspect ratio'. The value of the scaling factor can be interpolated from a table, which lists the typical values of κ from 0.3 to 0.5 and of aspect ratio from 0.2 to 8 (Hayes et al. 1972). When a/h is large, κ tends to be large; when a/h approaches 0, κ tends to be approaching the lower limit 1, when Equation 3.2 becomes the same as Equation 3.1.

Therefore, when the indenter radius is not much smaller than the initial thickness of the tissue, the effect of limited tissue thickness should be considered. The tissue thickness can be measured using various methods. In articular cartilage measurement, common methods include the needle penetration method (Jurvelin et al. 1995; Shepherd and Seedhom 1999), micrometer or caliper measurement (Mathiesen et al. 2004) and histological measurement. The disadvantages of these methods are that they can be used only for *in vitro* studies. Noninvasive measurement methods should be developed for *in vivo* tissue studies. Bader and Bowker (1983) used the skin caliper to measure tissue thickness; however, what they measured was the skin thickness rather than the thickness of the entire soft tissue layer. Modern medical imaging method such as x-ray radiography and magnetic resonance imaging (MRI) can be used for the measurement of soft tissue thickness (Silver-Thorn 1999; Gefen et al. 2001a,b). In most cases, the initial tissue thickness and deformation during the indentation test need to be measured separately, which is time consuming and inconvenient. Therefore, it is necessary to develop a technique that can measure both parameters simultaneously. This type of novel indentation technique is described in Chapter 4.

The other parameter that affects the results in calculating the tissue stiffness is Poisson's ratio, especially when the indenter radius is comparable to or larger than the tissue thickness. Poisson's ratio can be measured directly from an experimental test in which a regularly shaped tissue is compressed or extended and which is then calculated as the ratio of cross-sectional expansion and axial length change (Jurvelin et al. 1997). Indirect methods for measurement of Poisson's ratio are based on theoretical analysis of some mechanical tests. Jurvelin et al. (1997) proposed the use of the relationship between Young's modulus and aggregate modulus in unconfined and confined compression for the calculation of Poisson's ratio. The relationship as disclosed in Equation 3.2 between the Young's modulus measured by unconfined compression and the force/deformation as obtained using indentation test can also be used to calculate Poisson's ratio (Korhonen et al. 2002). Jin and Lewis proposed the measurement of Poisson's ratio using two indenters with different sizes (Jin and Lewis 2004). The relationship was obtained with infinitesimal deformation, and later Choi and Zheng further modified the method under the situation of finite deformation (Choi and Zheng 2005; Choi 2009). The latter group further reported a method to measure Poisson's ratio using a single indentation together with nonlinear effects (Zheng et al. 2009). When finite deformation is induced, it can be demonstrated through FEA that the relationship between the scaling factors κ and ν, a/h and d/h can be used to calculate Poisson's ratio, providing possibly an easy way of measurement. Through a biphasic indentation test, Poisson's ratio can also be evaluated, as introduced in the next subsection. In biological soft tissue analysis, Poisson's ratio is commonly assumed to be a constant to replace the complicated measurement method. Possible values are adopted in the range 0.45–0.5, assuming that soft tissues are nearly incompressible. With this assumption, a small fluctuation of Poisson's ratio is considered to have negligible effect on the measurement results. However, under specific conditions such as the soft tissues with restricted movement in a prosthesis, the stress analysis becomes very sensitive to the variation of Poisson's ratio and it needs special consideration in these circumstances. Another special situation is when soft tissues are deformed very slowly or the measurement is taken at an equilibrium state, thus allowing sufficient time for the water content to move out of the tissues. Under this situation, the tissues are compressible, leading to a relatively low Poisson's ratio. Jurvelin et al. (1997) reported that Poisson's ratio of bovine humeral head articular cartilage under equilibrium state was 0.185 ± 0.065, which indicated the cartilage was quite compressible in this situation.

3.2.1.2 Effect of Finite Deformation

The Hayes formula is obtained under the situation of infinitesimal deformation. In a practical test, the deformation may be up to 10%–30%, in which case the Hayes formula cannot be directly used because of either the finite deformation or the nonlinear tissue elasticity. A simple method to solve this question is to calculate Young's modulus in an incremental way (Zheng 1997):

$$E(i) = \frac{1 - \nu^2}{2a\kappa\left(\nu, a/(h - d(i))\right)} \cdot \frac{F(i+1) - F(i)}{d(i+1) - d(i)} \tag{3.3}$$

where i represents the ith sample point in the indentation. The sampling rate is assumed to be high enough so that the step from i to $i + 1$ can be taken to be quite small. The scaling factor κ is modified in the calculation to compensate for the change of tissue thickness from continuous deformation. To solve this problem better, Zhang et al. added another independent factor d/h inside the scaling factor, so that the equation becomes:

$$E = \frac{1-v^2}{2a\kappa(v, a/h, d/h)} \cdot \frac{F}{d} \tag{3.4}$$

where the scaling factor κ is related not only to Poisson's ratio and the aspect ratio but also to the deformation ratio d/h. The new scaling factor can be retrieved from a new table, with v, a/h and d/h as the independent variables (Zhang et al. 1997). Finite element study shows that κ nearly increases linearly with d/h. Especially, when v approximates 0.5 and a/h is large, the slope of the change with d/h is large. It should be noted that the nonlinearity between F and d is not necessarily induced by the nonlinearity of material properties or large deformation, because their relationship is also affected by the indenter's shape. For example, when a hemispherical indenter is used, the relationship between force and deformation at an infinitesimal deformation is (Jachowicz et al. 2007; McKee et al. 2011) given by

$$F = \frac{4R^{1/2}E}{3\left(1-v^2\right)} d^{3/2} \tag{3.5}$$

where
  R is the radius of the hemispherical indenter
  E is Young's modulus of the indented material.

It is clear from this equation that even when the deformation is small, the force and deformation is nonlinear with an exponential nature with exponent 1.5. Furthermore, when the substrate is not infinite but with a limited curvature, the situation would become more complicated as disclosed in a specific FEA (Lu and Zheng 2004). Further description on this can be found in Section 3.2.3.

3.2.2 Biphasic Analysis of Indentation Test

The single-phase indentation model is useful to describe the mechanical behaviour in the instantaneous state of a test. However, situations may exist where there is a large difference between the stiffness values for the same tissue in two different states, especially when there is obvious viscosity for the tested tissue. For example, the instantaneous modulus for the cartilage on the medial femoral condyle could be as high as 10.8 MPa; however, the equilibrium modulus was about 1.4 MPa, which caused some difficulty in explaining the test results (Julkunen et al. 2009). Therefore, in order to have a more effective analysis of the cartilage behaviour, Kuei et al. (1978) proposed a biphasic rheological model (called KLM biphasic model) for describing the mechanical behaviour of articular cartilage under indentation. The biphasic model assumes the first phase to be a porous solid matrix phase with isotropic and homogeneous linear elastic properties. Two parameters, namely the aggregate modulus H_A and Poisson's ratio ν, can be used to represent the material properties of the solid phase. The second phase is related to the interstitial fluid with a characteristic parameter of permeability k, representing the ease of motion of the fluid in the porous basal material. Mow et al. (1980) analysed the biphasic model using a confined compression test of articular cartilage. From the creep test, the aggregate modulus H_A as well as the permeability k can be obtained using a regression method, but not Poisson's ratio ν. Furthermore, the confined compression is not easy to perform because of the requirement for a precise preparation of tissue sample with respect to the container. Later, Mak et al. (1987) extended this model to the indentation test of articular cartilage. A simple numerical regression method was further developed to extract three different parameters, i.e. H_A, k and ν, from the creep indentation test (Mow et al. 1989), and this method has been widely adopted to study the mechanical properties of articular cartilage. The details of the numerical analysis method are described in the following.

It is assumed that a cylindrical porous rigid indenter is used in the test. The porous indenter allows the fluid of the cartilage to be transported freely from the indenter side. For the indentation test, it is assumed that μ_s and ν_s are the shear modulus and Poisson's ratio of the solid matrix, respectively, k is the permeability of the fluid, a is the indenter radius, h is the initial tissue thickness, F_0 is the constant indentation force for the creep test and d is the deformation. According to the biphasic indentation theory (Mak et al. 1987), the solution of the creep indentation test is totally defined by four parameters: Poisson's ratio ν_s, the indentation factor $F_0/(2\mu_s a^2)$, the time scaled factor $a^2/(H_A k)$ and the aspect ratio a/h. A fast numerical solution can be obtained by comparing the experimental deformation behaviour with the so-called master solutions. A master solution is defined as a solution of $d(t)$ when $F_0/(2\mu_s a^2)$ and a/h are defined from the experimental condition with $a^2/(H_A k) = 1$ and certain v_s which varies from 0 to 0.5. The shear modulus μ_s is calculated by the equilibrium state of the indentation test with a pre-determined v_s, where a master solution is defined:

$$\mu_s = \frac{1-v_s}{4a\kappa\left(v_s, a/h\right)} \frac{F_0}{d(\infty)} \tag{3.6}$$

where $d(\infty)$ represents the equilibrium deformation of the tissue under the creep test. When the master solutions are obtained, the experimental curves are compared with those master solutions by adding an extra time S in the $\log(t)$ axis. Assuming that the master solution with a time shift S is a function of $f\left(v_s, S, \log(t)\right)$ and the experimental curve is $d_{exp}\left(\log(t)\right)$, the optimal parameter (v_s, S) can be obtained by minimising the following criterion Q:

$$Q = \sum_{t_j} \left(\frac{f\left(v_s, S, \log(t)\right) - d_{exp}\left(\log(t)\right)/h}{d_{exp}\left(\log(t)\right)/h} \right)^2 \tag{3.7}$$

where t_j is the discrete time points used for the calculation. After obtaining the optimal Poisson's ratio v_s, the aggregate modulus of the tissue is calculated as

$$H_A = \mu_s \frac{2(1-v_s)}{1-2v_s} \tag{3.8}$$

In Equation 3.7, the time-scaled factor is $S = \log_{10}\left(a^2/kH_A\right)$ and, therefore, the permeability factor k can be calculated by

$$k = 10^{-S} \frac{a^2}{H_A} \tag{3.9}$$

Through calculations, it can be seen that using the creep indentation test, the three material parameters can be obtained simultaneously and it is not necessary to use a hypothesized Poisson's ratio for the calculation. The biphasic model has been widely used for the study of the mechanical properties of the articular cartilage. For example, it has been used to compare the difference of mechanical properties of the articular cartilage among different animal species (Athanasiou et al. 1991, 1995), the difference of cartilages at two contact sides in a joint (Athanasiou et al. 1994) and determine the spatial distribution of the cartilage stiffness in the joint (Roemhildt et al. 2006).

3.2.3 Indentation on Tissues with Curved Substrate

In the classic indentation model for soft tissues (Hayes et al. 1972), the tissue layer being tested is assumed to be flat and rigid. However, this condition cannot be always fulfilled for *in vivo* testing, such as indentation tests on limbs, which have a bony substrate with various curvatures. Lu and Zheng (2004) tackled this question using a series of finite element models of indentation with different curvatures of a bony substrate (Figure 3.6). The results demonstrated that the errors of the calculated Young's modulus using a finite deformation indentation model (Zhang et al. 1997) are closely related to the ratio between the indenter radius and substrate curvature as well as the ratio between the indenter radius and tissue thickness. The error can be up to 30% when the ratio between the indenter radius and substrate curvature is 0.375 and the ratio between the indenter radius and tissue thickness is 2.0, and up to 10% when the two ratios are 0.375 and 0.2, respectively (Lu and Zheng 2004). Therefore, when conducting indentation tests on tissues with a curved substrate, such as limb tissues, the results should be carefully interpreted, and

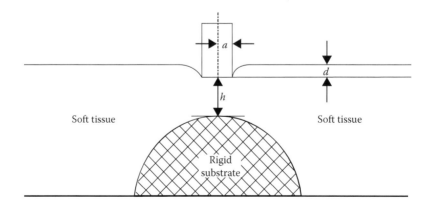

FIGURE 3.6 Illustration of an indentation test on tissues with curved substrate.

necessary corrections made in accordance with the results provided by Lu and Zheng (2004) should be considered.

3.2.4 Indentation on Tissues with Curved Surface

Most of tissues can have a flat surface for the indentation test. However, when the indentation test is used for the assessment of corneal elasticity, special consideration should be given as the cornea inherently has a curvature (Figure 3.7). The conventional indentation models cannot be directly applied for testing the cornea. Ko et al. (2013) reported a model to extract Young's modulus from the indentation data of cornea under a certain value of IOP, called tangent modulus, which may vary under different IOP values due to the nonlinear biomechanical properties of the cornea. The tangent modulus can be calculated using the following equation (Ko et al. 2013):

$$E_{IOP} = \frac{k(R-h/2)\sqrt{1-v^2}}{h^2} \left. \frac{dF}{d\delta} \right|_{IOP} \tag{3.10}$$

where
  E_{IOP} is the tangent Young's modulus obtained under a constant IOP
  F is the force applied on the indenter
  δ is the indentation depth
  R is the curvature of the cornea
  h is the cornea thickness
  v is Poisson's ratio
  k is a geometry constant determined from μ using a table (Young 1989; Ko et al. 2013):

$$\mu = a\left[\frac{12(1-v^2)}{(R-h/2)^2 h^2} \right]^{1/4} \tag{3.11}$$

Recently, there has been increasing interest in measuring the elasticity of the cornea because it not only affects the IOP measurement but also relates to different kinds of eye diseases. Some other methods for corneal elasticity measurement such as the air puff test will be introduced in Chapter 4.

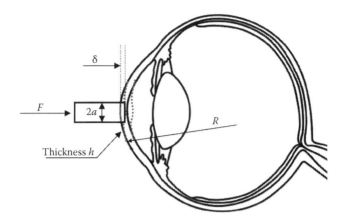

FIGURE 3.7   Illustration of the indentation test on cornea to measure its elasticity.

3.2.5 Nonlinear Viscoelasticity and Finite Element Analysis

Because of the complexity of constitution of the soft tissue, the models that have been introduced in the previous subsections are not enough for practical applications. For example, the biphasic model can explain the confined compression of soft tissue well, whereas the behaviour of soft tissue in unconfined compression is not as well described because this model does not consider the intrinsic viscoelastic properties of the solid matrix in biological soft tissues. Therefore, it is specifically necessary to add more material properties to the tissue model. Meanwhile, these complicated tissue models may make the theoretical analysis difficult, and then a numerical method is a more appropriate method in these situations. The related studies are scattered in the literature, and a separate chapter on FEA is provided in this book. More details can be found in Chapter 9.

3.3 ENDING REMARKS

In general, the indentation test can be easily used for both *in vitro* and *in vivo* studies because   no special specimen preparation is needed. When the overall deformation is controlled within a certain range and the time of loading is limited, it is safe and harmless to the tested object. Therefore, it has been widely used for the noninvasive measurement of soft tissue elasticity. The measured elasticity parameters can be used for computer simulation, diagnosis of pathology and evaluation of treatment efficiency. This chapter mainly reviews the conventional indentation test and its analysis. In most systems, the displacement of the indenter tip is used as an indicator of the deformation, and the tissue thickness needs to be measured using another technique. Based on different application circumstances, different indentation systems have been developed. With fast advances in technology, these systems get better and better in aspects such as portability, operation ease, measurement reliability and repeatability. New indentation techniques are also emerging in this field such as ultrasound indentation and nanoindentation, as described in the next chapter. However, in the elasticity measurement of soft tissues, indentation is also facing some challenges from some other test methods, such as suction/aspiration (Diridollou et al. 2000; Nava et al. 2008), shear wave propagation method or transient elastography (Sandrin et al. 2003). We feel that indentation will still have some applications in the measurement of soft tissue elasticity, and may serve as a reference method for developing new techniques. Future direction is to develop specific indentation systems in different application fields to meet the specific requirements in that field and then to standardise the test procedure and method for extraction of tissue elastic properties, for quantitative and objective study both in laboratory research and clinical applications.

.

CHAPTER 4

Novel Indentation Measurement of Tissue Elasticity *in Vivo*

4.1 INTRODUCTION

In Chapter 3, the concept of traditional indentation is introduced, and traditional indentation utilising various techniques to measure elasticity parameters has also been reviewed thoroughly. In general, two parameters need to be recorded in an indentation test: force and deformation (tissue initial thickness is another parameter normally required in calculation). The force can be measured by a load cell (Mak et al. 1994), a strain gauge (Lyyra et al. 1995), a fibre-optic sensor (indirectly) (Komi et al. 1996) or a photoelastic sensor (Gefen et al. 2001a). In the case of deformation, traditionally it is measured by quantifying the displacement of the indenter head. Instruments that can be used for displacement measurement include a linear variable differential transformer (LVDT) (Mak et al. 1994; Vannah and Childress 1996; Rome and Webb 2000), a potentiometer (Bader and Bowker 1983; Xiong et al. 2010), an electromagnetic spatial sensor (Vannah et al. 1999; Lu et al. 2009c) or a mechanical 3D location system (Klaesner et al. 2001). Traditional techniques for deformation measurement have some obvious disadvantages. For example, the indentation systems with LVDT, potentiometer or a mechanical arm are not quite portable; electromagnetic spatial sensors are vulnerable to interferences from the surroundings. With the advancement of medical imaging technologies, researchers also adopted such modern methods for the measurement of tissue thickness and deformation. Armstrong et al. (1979) utilised the x-ray radiography to observe the change of cartilage *in vitro* in static loading. Gefen et al. (2001b) adopted a digital radiographic fluoroscopic system to measure the deformation of the plantar soft tissues in gate. However, the x-ray imaging system is too large to be used in portable device, and also has the risk of radiation-induced side effects. MRI can also be used for the measurement of thickness and deformation, with good contrast of various soft tissues (Gefen et al. 2001a). The problem comes from the very high cost for MRI detection. Also, the system profile is too large to be used for the development of a portable instrument. Ultrasound, which utilises the wave propagation induced by small vibrations, is a mechanical means that can be conveniently used for the measurement of tissue thickness and deformation. Therefore, ultrasound can be incorporated in the traditional indentation system as a new modality to measure the test information, which is generally called as 'ultrasound indentation'. Although an ultrasound machine seems to be large in the clinical field, a small-profile single or array-based probe can be incorporated in the indentation system design, which makes the system portable enough for the ease of operation. Zheng and Mak first designed a pen-sized ultrasound indentation probe, which was successfully applied for studying various soft tissues and the related pathologies for tissue characterisation (Zheng and Mak 1999; Zheng et al. 1999b, 2000a,b, 2012; Leung et al. 2002; Huang et al. 2005; Lau et al. 2005; Makhsous et al. 2008; Kwan et al. 2010; Siu et al. 2010; Chao et al. 2011a). In order to facilitate the use of a high-frequency ultrasound transducer and realise fast elasticity mapping, fluid jet indentation methods including water-jet and air-jet indentation were then developed, in which the non-contact indentation methods have their unique advantages in some specific applications. Furthermore, optical imaging methods have also been incorporated in the indentation technique. For example, optical coherence tomography (OCT) has been used in the indentation for the non-contact measurement of tissue deformation (Huang et al. 2009). As fibre-optic techniques can be used for the optical system design and signal transmission, an indentation system utilising the optical technique has the potential to be realised in a portable format or in endoscopic applications. For example, the newly developed OCT technique has been successfully designed for use in arthroscopic (Pan et al. 2003) and gastrointestinal endoscopic channel (Izatt et al. 1996) for a better diagnosis of joint and digestive tract diseases. Another advantage of the use of acoustic or optical detection in an indentation test is that some other material properties such as the acoustic or optical properties and morphological characteristics of the tissues can be measured simultaneously. For example, available acoustic properties include the speed of sound, the surface reflection, attenuation and backscattering coefficient. For optical signals, the parameters include the refractive index, optical surface reflection, internal scattering and optical attenuation. The morphological parameters include the thickness and the surface roughness of the tissue. These properties, together with the mechanical properties obtained from the indentation test, can be used to characterise the tissue status in a multi-modality way (Wang et al. 2010b). This will be further discussed in Section 4.2.4.

On the other hand, the structural complexity in biological soft tissues has made the mechanical properties of soft tissue heterogeneous and anisotropic. If the mechanical properties of soft tissues can be measured microscopically, then the relationship between function, structure and material properties can be understood in a better way, which would advance the basic research in disease pathology, tissue repair and tissue engineering. In certain situations, the dimension of the tissue, such as the articular cartilage in a mouse model, which has only a thickness of several hundred micrometers, is also limited. Therefore, the spatial resolution of measurement should be increased correspondingly. In order to measure the mechanical properties of soft tissues on a microscopic scale, the nanoindentation technique in material testing has also been introduced in this field. Based on the difference of scale for measurable deformation, the general concept of nanoindentation includes the instrumented indentation and the atomic force microscopy-enabled indentation. They have different applications in different fields of biological measurement and have their own advantages and disadvantages in applications (Van Vliet 2011). In biological applications, the nanoindentation was first used in the elasticity measurement of two mineralised hard tissues, i.e. bone (Rho et al. 1997; Turner et al. 1999; Ferguson et al. 2003; Haque 2003) and dental tissue (Habelitz et al. 2001; Cuy et al. 2002; Kinney et al. 2003; Angker and Swain 2006). These two materials received broad application of nanoindentation because they, compared to soft tissues, are more similar to traditional engineering materials such as metals or ceramics. Therefore, the measurement range in force and deformation is comparable to that of engineering materials, and it is relatively easier to adopt the nanoindentation instrument for these two hard tissues. Furthermore, these two tissues have an obvious layered structure (osteons and lamellae in bone; dentine, enamel and cement in dental tissue), which makes the measurement of localised elasticity more meaningful. However, a nanoindentation test of biological tissues is easily affected by a large number of factors, such as the condition of hydration, the surface preparation method (polished or unpolished), the storage medium, duration of storage and the loading frequency (static, quasi-static or dynamic) (Lewis and Nyman 2008). Therefore, the method of nanoindentation of industrial materials should be modified for testing biological tissues. Based on successful experience in nanoindentation of hard tissues, it is further adopted for the measurement of the interface between hard and soft tissues. Gupte et al. (2005) utilised nanoindentation to measure the elasticity at the interface of bone and calcified cartilage and its relationship with the content of mineralisation. It was found the relationship between stiffness and mineral content was different for bone and the calcified cartilage. Among the various soft tissues, cartilage has been one of the most frequently reported tissues where nanoindentation was used for the measurement of its mechanical properties (Ebenstein et al. 2004; Simha et al. 2004, 2007; Li et al. 2006; Franke et al. 2007, 2011). Nanoindentation has also been reported in the study of elasticity in the stratum corneum (Yuan and Verma 2006) and vascular tissues (Ebenstein et al. 2009). The factors affecting the nanoindentation measurement of soft tissues are different from those in the case of hard tissues; for example, it is affected by the shape of the indenter and the adhesion between the indenter and soft tissues (Ebenstein and Pruitt 2006), which will be introduced in Section 4.3.

4.2 NOVEL INDENTATION MEASUREMENT

4.2.1 Methods

4.2.1.1 Ultrasound Indentation

As seen from the introduction of traditional indentation systems in Chapter 3, the development generally is on several fronts: from large to portable in size, from complicated to ease of operation and from manual to automatic in data collection. However, one problem that has not yet been solved is the simultaneous measurement of thickness and deformation, which can be done when ultrasound indentation is used. Zheng and Mak (1996) utilised a single-element ultrasound transducer operating at 5 MHz to design a portable indentation probe, which was called an 'ultrasound indentation' or a 'tissue ultrasound palpation system' (TUPS), to measure the stiffness of soft tissues (Figures 4.1 and 4.2). A plane-ended, cylindrical, unfocused ultrasound transducer with a central frequency

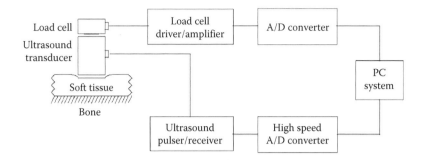

FIGURE 4.1 Diagram of an ultrasound indentation system. (From Zheng, Y.P. and Mak, A.F.T., *IEEE Trans. Biomed. Eng.*, 43, 912, 1996.)

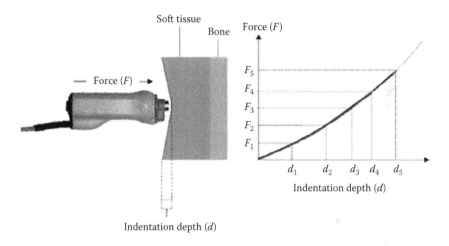

FIGURE 4.2 Illustration of the indentation test using an ultrasound indentation probe.

of 5 MHz was installed at the tip of a pen-sized probe. A force sensor was connected in series with the ultrasound transducer for the measurement of indentation force. The ultrasound transducer was connected to an ultrasound pulser/receiver for the generation and receiving of ultrasound signals, and the force sensor was connected to an amplifier for the amplification of the sensed force signals. The two types of signals were then input to a computer for the synchronised sampling, display and storage (Figure 4.3). The measurement of deformation and thickness was completed by tracking the echoes reflected from tissue interfaces. Multiple echoes can also be tracked using the software so that the deformation of individual tissue layers can be obtained (Figure 4.4). Because of its small size, the TUPS probe can be used for manual operation and

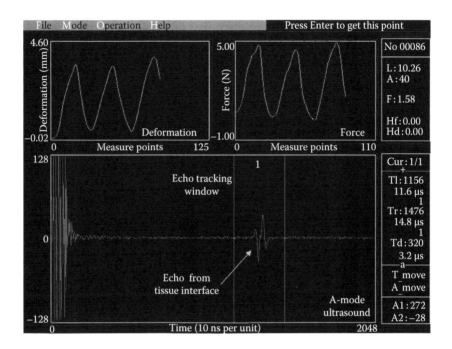

FIGURE 4.3   Software interface for tissue ultrasound palpation system (TUPS). The deformation of tissue is calculated from the movement of echo in A-mode ultrasound signal, according to a pre-determined speed of ultrasound, commonly using 1540 m/s in most soft tissues.

the whole device is also portable (Figures 4.5 and 4.6). In comparison with most conventional method of measuring tissue deformation under indentation, ultrasound indentation provides a unique advantage of self-referencing, i.e. the measured deformation will not be affected, when the tested body part moves during the measurement. For other methods, such as LVDT, they use the displacement of the body surface as reference, which can cause errors in tissue deformation measurement when the body moves. This unique feature, together with the small-sized probe, makes TUPS very popular for the assessment of various tissues, and a large number of research papers have been published, including foot plantar tissues, neck tissue fibrosis induced by radiotherapy (Huang 2005), buttock tissues developing ulcers, breast tissue, articular cartilage, muscle with different contraction levels, finger tips, scar tissue, etc. Some of these applications will be introduced in more detail in Chapter 10.

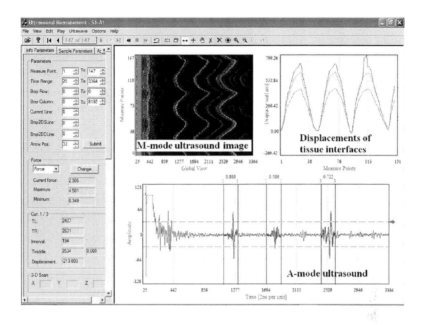

FIGURE 4.4   TUPS software used to automatically track the movement of multiple echoes in A-line ultrasound signals, and thus to measure deformation of individual tissue layers under indentation.

FIGURE 4.5   Portable version of TUPS with a finger-sized probe for tissue thickness and elasticity measurement.

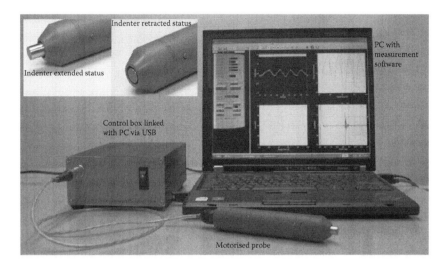

FIGURE 4.6 PC-based TUPS device with a motorised probe for the measurement of tissue thickness and elasticity. (Inset) The motorised probe with the indenter in extended and retracted states.

In practical operation, the pre-loading can be adjusted by a manually applied force, and the maximum deformation can be controlled using the feedback from the echo movement. To obtain an optimised signal from the tissue interface, adjustment of the probe orientation is needed to get the maximum reflection signal when the transducer is aligned perpendicular to the tissue surface. Zheng et al. (1999b) demonstrated that the magnitude of ultrasound echo is very sensitive to the change of probe alignment, and thus can be used as a good indication of misalignment (Figure 4.7). To reduce the effects of pre-conditioning, 3–5 cycles of loading and unloading are usually needed before the real data collection to achieve reliable results from the measurement. Because of the viscosity of soft tissues, the tissue behaviour is normally affected by the indentation speed. Therefore, it is necessary to control the indentation speed. An indentation speed with which the operator feels comfortable for the test is suggested, and reliable measurement results can be achieved for the specific operator.

Li et al. (2004, 2005, 2006) proposed an ultrasound indentation system that was similar to TUPS. A dual-element ultrasound transducer was used for the design. The transducer was divided into a left and a right part, which were responsible for the emission and receiving of ultrasound signals, respectively. A load cell was connected with the ultrasound transducer for the measurement of force during indentation. The combination of ultrasound transducer

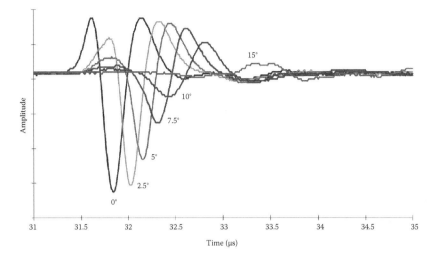

FIGURE 4.7   Change of ultrasound echo reflected from the bottom interface of a fresh porcine tissue under different degrees of misalignment of TUPS probe.

and load cell was enclosed in an aluminium cover for easy operation. Suh et al. (2001) proposed an ultrasound indentation method to measure the speed of sound in articular cartilage. This method fixed the ultrasound transducer in a micrometer and used it to indent the cartilage. The indentation depth could be measured from the micrometer, and the change of propagation time between transducer and bone could be measured from the ultrasound signals. Utilising these two signals, the speed of sound in cartilage, which had the potential to detect early degeneration of the articular cartilage, could be measured (Suh et al. 2001). However, as it is not so possible to install a fixed frame with accurate measurement of indentation for *in vivo* studies, this method is applicable only to *in vitro* measurements.

Laasanen et al. (2002) embedded the ultrasound indentation system into arthroscopy for the measurement of articular cartilage properties. They installed a 10 MHz ultrasound transducer with a diameter of 3 mm on the tip of an arthroscope, which could be directly used to indent the cartilage. It should be noted that additional acoustical properties of the cartilage can also be measured from the ultrasound transducer. However, it is necessary to insert a bolster between the transducer surface and the cartilage because a certain spatial distance is needed to obtain the surface acoustic reflection from the cartilage. If no space is set between the transducer and the cartilage, the reflection from the ultrasound surface and cartilage surface, which cannot be separated, will overlap. The insertion of an extra bolster will be inconvenient during the measurement, which may hinder its broad application *in vivo*. On the other hand, if thin tissues need to be measured by ultrasound, higher spatial resolution, which can be achieved by increasing the frequency, is necessary. However, the surface is normally concave for a high-frequency ultrasound transducer, and therefore it is inappropriate to use the transducer itself directly as an indenter. Water-jet ultrasound indentation was proposed to tackle this challenge, which is introduced in Section 4.2.1.2.

The ultrasound indentation probe is normally operated manually, but some alternative configurations have also been developed. Figure 4.8 shows a device developed for the assessment of foot plantar tissue, where the ultrasound transducer was located on a platform to deform the tissue when the body weight was applied (Zheng et al. 2011). The same group further developed a system with a motorised ultrasound indentation probe together with a force sensor and cameras installed inside an assessment platform on which the subjects could stand to test their foot plantar tissues (Figure 4.9) (Zheng et al. 2012).

In addition to the one-dimensional ultrasound, two-dimensional ultrasound imaging can also be incorporated for the indentation test. Hsu et al. (1998) attached a push–pull scale with an ultrasound transducer of 7.5 MHz, in which the indentation was conducted through the vertical movement of

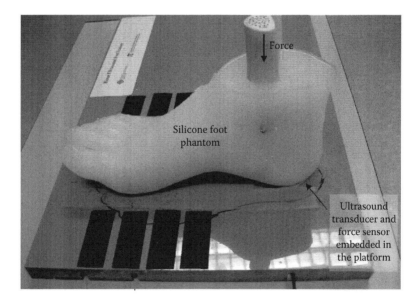

FIGURE 4.8 Platform embedded with an ultrasound transducer and a force sensor for foot plantar tissue biomechanical assessment.

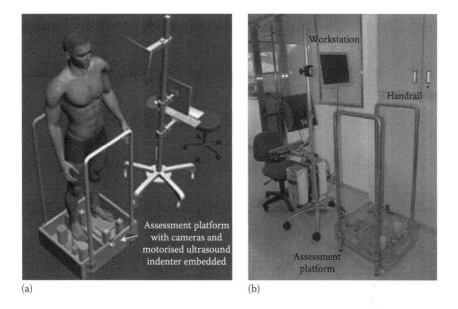

(a)   (b)

FIGURE 4.9 System for foot plantar tissue assessment with motorised ultrasound indenter. (a) 3D design of the system. (b) Photo of the actual system.

a hand knob. The indentation details could be obtained from the 2D ultrasound imaging, and this system was applied at the heel pad for the measurement of the mechanical properties of plantar tissues. Later, the system design was improved by the adoption of an electronic force sensor and driven by a motor for automatic operation (Hsu et al. 2009). Kawchuk et al. introduced indentation for the measurement of spine deformability (Kawchuk and Elliott 1998; Kawchuk et al. 2000, 2001a,b). As the change of spine deformability may be a direct reflection of its pathology such as degeneration, it is useful to measure the movement of the spine during an indentation process. This method utilises the subtraction of the ultrasound-measured deformation of soft tissues from the total displacement measured from the indenter tip by LVDT to measure the displacement of the bone during indentation. Han et al. (2003) adopted a six degree-of-freedom force sensor to measure the indentation force in an ultrasound indentation system. The transducer was installed on one side of the force sensor, and a hand-held holder was attached to the other side of the transducer for convenience of operation. A 2D ultrasound transducer was used in this indentation system for the measurement of tissue deformation. Figure 4.10 shows such a system, which was developed in the authors' group for the measurement of breast tissue elasticity and its change during a menstrual cycle (Li et al. 2008b, 2009).

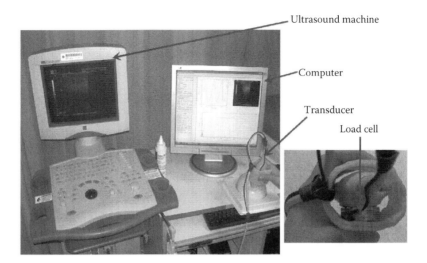

FIGURE 4.10   Use of a 2D ultrasound imaging system in indentation test. (Inset) A load cell installed with the 2D ultrasound probe for recording load during the indentation test.

4.2.1.2 Ultrasound Water-Jet Indentation

To achieve indentation using high-frequency ultrasound for measuring the elasticity of a thin tissue layer, Lu et al. (2005) proposed a novel water-jet ultrasound indentation method. This indentation system utilises a water jet to indent the tissue in a non-contact way (Figure 4.11) (Lu et al. 2005; Lu 2007). The indentation force can be adjusted by the water pressure, and the ultrasound signal is used to measure the thickness and deformation. Experiment on silicone phantoms showed a high correlation between the stiffness measured from water-jet indentation and from a compression test. Because of non-contact indentation, this method can also be used for a mapping of tissue elasticity, which was demonstrated in a phantom test (Lu et al. 2007). Ultrasound frequency of 20–50 MHz was used in water-jet ultrasound indentation systems. This technique has been successfully used for the evaluation of the patella–tendon junction healing in a rabbit model *in vitro* (Lu et al. 2009b). As shown in Figure 4.11d, the surface of the patella–tendon junction is very complicated, thus it is not possible to use any other method to map the elasticity distribution along such a rough surface. Water-jet ultrasound indentation provides a unique measurement approach for mapping the elasticity distribution of tissues with complicated surface conditions using its 'soft' scanning feature.

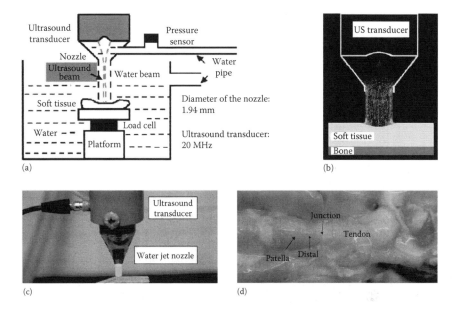

FIGURE 4.11 Water-jet ultrasound indentation. (a) Schematics showing water-jet indentation test on soft tissue samples. (b) Illustration of the enlarged water-jet ultrasound indentation probe. (c) Actual water-jet ultrasound indentation probe. (d) Patella–tendon junction tested using water-jet indentation for the elasticity mapping. (From Lu, M.H. et al., *Ann. Biomed. Eng.*, 37, 164, 2009.)

Huang and Zheng (2009, 2013) have further developed the water-jet ultrasound indentation technique into an endoscopic probe for the assessment of articular cartilage (Huang and Zheng 2009, 2013; Huang 2013). An arthroscopic guiding channel with a diameter of 5.5 mm was modified by including an intravascular ultrasound (IVUS) probe ($\Phi = 1$ mm, $f = 20$ MHz) and a water path (Figure 4.12). The water or saline ejects from a hole located at the tip of the probe, where the IVUS transducer is also installed. The water beam applies a load to the articular cartilage to cause deformation, which is detected by real-time imaging of the IVUS system. The prototype of the water-jet ultrasound indentation probe was used to measure the elasticity of articular cartilage of porcine knee joints *in situ* (Figure 4.13). Preliminary results demonstrated that the probe could differentiate cartilage before and after trypsin digestion (trypsin can damage proteoglycan in articular cartilage, thus mimicking degeneration). In addition to the measurement of tissue elasticity, the probe can also provide high-resolution ultrasound imaging and also ultrasound tissue characterisation with different acoustic parameters of articular cartilage.

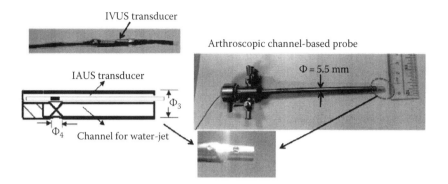

FIGURE 4.12 Endoscopic water-jet ultrasound indentation probe and its components, including an intravascular ultrasound transducer.

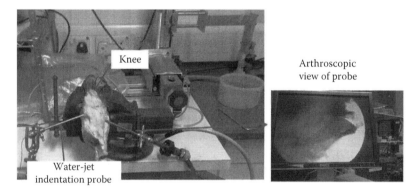

FIGURE 4.13 Experimental set-up of water-jet ultrasound indentation test for articular cartilage of porcine knee joints *in situ*. (Insert) Arthroscopic view of the porcine knee joint cavity and the probe.

4.2.1.3 Optical Indentation

In addition to ultrasound imaging, different optical imaging techniques have been applied in the field of mechanical tests, such as indentation to measure tissue deformation. The advantages of optical imaging in biological applications include the high spatial resolution and the avoidance of coupling media. However, high spatial resolution always comes with low penetration, and therefore only a superficial layer of the soft tissues can be viewed using optical imaging in the indentation test. Duda et al. (2004) reported a water-jet indentation system that utilised an optical signal for the measurement of the deformation of cartilage in an indentation test (Figure 4.14). Visible light was used in their system, and because its intensity was varied by the distance between the probe and the cartilage surface, the detection of intensity change

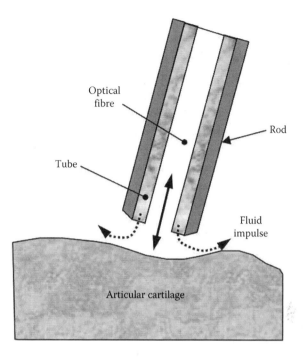

FIGURE 4.14 Illustration of water-jet optical indentation probe, which uses the light intensity change to detect the induced deformation of cartilage.

could be used as a means to measure cartilage deformation. The system was designed in an arthroscopic channel for the intra-articular operation, and the relationship between force and deformation could be measured from the system. One disadvantage of the system was that the detected optical signal could be affected by external factors such as the color of the cartilage surface and environmental background light. Liu et al. (2011) designed a rolling indentation probe with physical contact, which utilised the optical signal to detect the indentation force and deformation. In the centre of the probe was a rolling indenter attached with a spring system having a large elastic constant. The small deformation of the spring system could be detected by an optical sensor for the measurement of the indentation force. Around the central sensor, there were another four sliding rods, the movement of which could also be detected by optical sensors. As the detection of the optical signal was done inside the housing, the effect of external factors could be minimised. Recently, an improved version of the probe, in which a pneumatic system was incorporated into the system design, was reported. A central sphere, which was driven by a high-pressure airflow for the indentation, was used, and the surrounding sliding rods were designed to be closely attached to the skin surface for deformation measurement together with displacement information measured from the central sphere (Wanninayake et al. 2012).

Corneal biomechanical properties have been shown to be correlated with the pathologies of some corneal degeneration diseases like keratoconus, keratectasia, pellucid marginal degeneration (Andreassen et al. 1980; Ortiz et al. 2007) and glaucoma (Johnson et al. 2007). However, there is still difficulty in accurately measuring the corneal properties *in vivo* due to the lack of a proper measurement method. The Ocular Response Analyzer (ORA; Reichert, Corp., Buffalo, NY) is the first instrument that allows the evaluation of corneal biomechanical properties *in vivo*. It provides corneal biomechanical properties, including corneal hysteresis (CH) and corneal resistance factor (CRF), with CH quantifying the corneal viscoelastic mechanical damping ability and CRF indicating the whole corneal viscoelastic resistance (Figure 4.15) (Luce 2005).

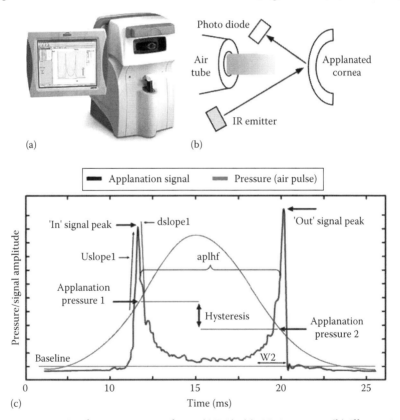

FIGURE 4.15 Ocular response analyzer (ORA). (a) ORA system. (b) Illustration of optical detection of applanation of cornea. (c) Signals of air pulse pressure and applanation and illustration for hysteresis measurement.

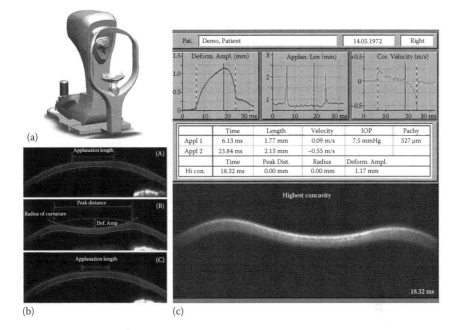

FIGURE 4.16 Corvis ST system for corneal mechanical measurement. (a) Corvis ST device. (b) Corneal images at different moments after an air pulse. (c) Measurement interface of Corvis ST.

While it gives the viscoelasticity of the cornea, the interpretation of ORA data is difficult. Its sensitivity and efficiency in identifying keratoconus suspects or evaluation of the cross-linking effect, the recent and most popular intervention to reverse keratoconus, were questioned by many (Luce 2005; Goldich et al. 2009; Fontes et al. 2011). Newer corneal visualisation with Scheimpflug Technology (Corvis ST; Oculus Inc., Wetzlar, Germany), based on the Scheimpflug imaging technique, has become another method for measuring corneal biomechanical properties in the clinic since 2010 (Figure 4.16). The corneal displacement amplitude during an air-puff indentation holds promise to yield relevant parameters related to the corneal biomechanical properties; however, at present, none of the parameters from Corvis ST can be treated as an intrinsic mechanical parameter (Lau and Pye 2011). Theoretically, the Scheimpflug imaging technique is limited by geometrical and optical distortions, which make careful correction necessary beyond the quantitative extraction of biomechanics-related information (Dubbelman et al. 2002; Rosales and Marcos 2009). Tian et al. (2014) reported more parameters correlated with keratoconic eyes using Corvis ST.

Recently, the fast development of optical coherence tomography (OCT) has attracted a great deal of attention from researchers for its potential applications in biomedical engineering. Compared to traditional ballistic imaging methods, such as confocal imaging and multi-photon imaging, the advantage of OCT comes from its relatively larger penetration depth (2–3 mm in turbid tissues) while achieving a relatively high spatial resolution (in the scale of micrometers). Yang et al. (2007) proposed to observe tissue deformation during indentation using OCT. In the experimental design, a spherical indenter was used to indent the sample (hydrogel phantom) from above, and then OCT was used to observe the change of sample deformation from the bottom. It should be noted that in this system design, the observation of the indentation was performed from the bottom of the sample, which is possible only for some thin tissue sample and not so practical for *in vivo* tissue testing.

A fluid jet can also be incorporated with optical imaging for the measurement of tissue elasticity. Because no extra coupling medium is needed for the optical imaging, it can be easily introduced in the indentation system. Huang et al. proposed an OCT-based air-jet indentation system that could be used to measure the mechanical properties of soft tissues (Huang et al. 2009) (Figure 4.17). In this system, an air jet was used as a substitute for the traditional contact indenter to compress the tissue, and OCT was employed to measure the displacement of the tissue surface caused

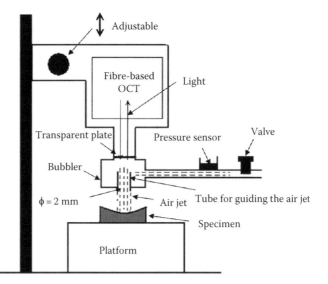

FIGURE 4.17 Illustration of optical coherence tomography-based air-jet indentation test on a tissue specimen.

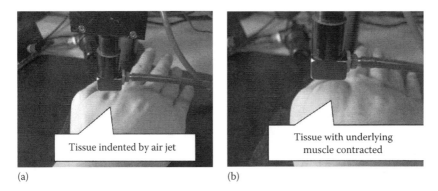

(a) (b)

FIGURE 4.18   Optical coherence tomography -based air-jet indentation test on hand tissues *in vivo*. (a) Tissue is indented by the air jet when it is soft. (b) Tissue is much more difficult to be indented when the underlying muscle is contracted.

by compression (Figure 4.18). The indentation force induced on the tissue could be calculated from the air pressure measured inside the air pipe. The relationship between the indentation force and the air pressure could be obtained from a pre-test calibration process.

Because of its non-contact feature and accurate measurement of tissue deformation, OCT-based air-jet indentation has been applied for measuring the elasticity of various tissues. Huang et al. (2011b) used it to quantify the stiffness change in degenerated articular cartilage, and Cheing and her team investigated the biomechanical properties of wound tissues in diabetic foot (Figure 4.19) (Chao et al. 2010, 2011b) and rat wound models

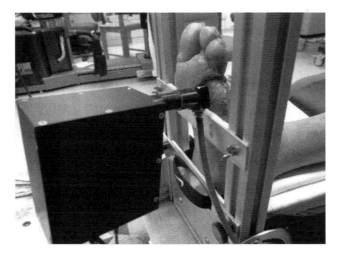

FIGURE 4.19   Optical coherence tomography-based air-jet indentation test on diabetic foot to study ulcer healing.

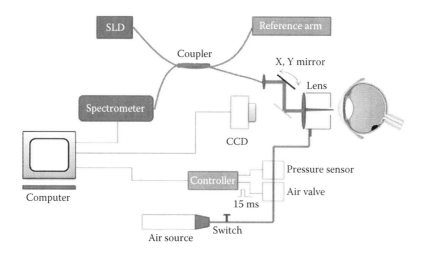

FIGURE 4.20 Block diagram of optical coherence tomography-based air-jet indentation system for assessment of corneal biomechanical properties.

(Chao et al. 2013). The system has also been used for the measurement of cornea elasticity (Zheng et al. 2014; Wang et al. 2015). In comparison with other applications, the system for cornea test can provide transient indentation using a short-duration air puff (15 ms) and 20,000 frames of A-mode signals (A-lines) per second (Figures 4.20 and 4.21).

4.2.2 Indentation Analysis

The analysis of ultrasound indentation is similar to that of traditional contact indentation with a rigid indenter. For a cylindrical indenter, when the indentation force deformation and the initial thickness of the tissue are obtained, the following equation can be used to calculate the mean Young's modulus (E) of the tissue (Hayes et al. 1972):

$$E = \frac{1-v^2}{2a\kappa(v, a/h)} \cdot \frac{F}{d} \tag{4.1}$$

where
    v is Poisson's ratio
    a is the radius of the cylindrical indenter
    F is the indentation force
    d is the indentation deformation
    κ is a scaling factor related to v and the aspect ratio a/h

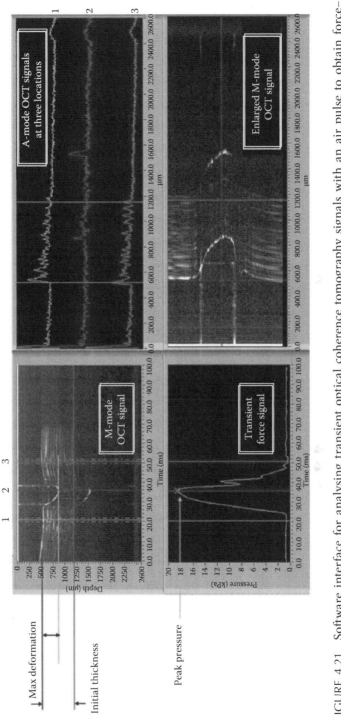

FIGURE 4.21   Software interface for analysing transient optical coherence tomography signals with an air pulse to obtain force–deformation relationship of cornea. (From Wang, L.K. et al., *Opt. Precis. Eng.*, 23, 325, 2015.)

The ratio of F/d is obtained from the linear regression of the indentation force and deformation. When there is large deformation, this equation is modified as (Zhang et al. 1997)

$$E = \frac{1-v^2}{2a\kappa\left(v, a/h, d/h\right)} \cdot \frac{F}{d} \tag{4.2}$$

where an additional factor d/h is added to the scaling factor for compensating the effect from the indentation nonlinearity. Other models of finite strain theory, such as the hyper-elastic model (Samani and Plewes 2004), can also be incorporated into the indentation model to obtain the mechanical properties of tissues, which normally are the optimised coefficients of the strain energy function from the measured indentation curve.

When the viscosity is considered, further modelling is necessary to study the mechanical behaviour of the soft tissues. The spring and dashpot model, introduced by Fung (1993a), can be used to describe the viscoelastic behaviour of the soft tissues. Alternatively, a viscoelastic model such as based on the quasilinear viscoelasticity (QLV) theory can be adopted for the extraction of mechanical properties of the soft tissue. Zheng et al. proposed the following QLV model combined with indentation analysis for extracting the viscoelastic properties of the soft tissues (Fung 1993c; Zheng and Mak 1999):

$$F(t) = F^{(e)}\left(u(t)\right) + \int_0^t F^{(e)}\left(u(t-\xi)\right)\frac{\partial G(\xi)}{\partial \xi}d\xi \tag{4.3}$$

where
  $u(t) = w(t)/h$ is the relative deformation ratio with respect to the initial thickness of the tissue
  $F^{(e)}(u)$ is the instantaneous force response when the indentation is u
  $G(t)$ is a force relaxation function along with time

The form of the function F can be chosen as a polynomial of u, such as (Zheng and Mak 1996)

$$F^{(e)}\left(u\right) = C_1 u + C_1 u^2 + C_3 u^3 \tag{4.4}$$

where C_i is the constant coefficient for the polynomial. For a contact indentation test, this equation can be written as follows (Hayes et al. 1972; Zhang et al. 1997; Zheng and Mak 1999):

$$F^{(e)}(u) = \frac{2ah\kappa(v,a,h,u)}{1-v^2} E^{(e)}(u) \qquad (4.5)$$

where the elastic modulus can be supposed to be a polynomial function of the deformation, for example:

$$E^{(e)}(u) = E_0 + E_1 u \qquad (4.6)$$

where

E_0 is the initial modulus

E_1 is the nonlinear factor, which is linearly dependent on the deformation ratio

The force relaxation function can be written as

$$G(t) = 1 - \alpha + \alpha e^{-t/\tau} \qquad (4.7)$$

where

α is a viscosity-related constant

τ is a time constant

According to the deformation history of $u(t)$ and the initial thickness of the soft tissue, if the model parameters (E_0, E_1, α and τ) are given, the force response $F_s(t)$ of the indentation can be modelled using the QLV model, which can be compared with the measured experimental force $F_e(t)$. A simulation error factor can be calculated as

$$S_{err} = \sqrt{\frac{\int_0^T \left(F_s(t) - F_e(t)\right)^2 dt}{\int_0^T F_e^2(t) dt}} \qquad (4.8)$$

In a discrete signal form, the force response at any discrete time i using the indentation history $u(j)$ ($0 < j < i$) can be written as (Huang et al. 2005):

$$F(i) = \frac{2ah}{1-v^2}\left[\kappa(u(i))\left[E_0 u(i) + E_1 u^2(i) \right] \right.$$

$$\left. - \frac{\alpha}{\tau} \sum_{j=0}^{i} \kappa(u(i-j))\left[E_0 u(i-j) + E_1 u^2(i-j) \right] e^{-j \cdot \Delta t / \tau} \Delta t \right] \quad (4.9)$$

where Δt is the time interval between two adjacent data points, and E_0, E_1, α and τ are defined as before. The scaling factor $\kappa(u(i))$ for an arbitrary $u(i)$ can be interpolated from the indentation-dependent values provided by Zhang et al. (1997), and used to compensate the nonlinear effect caused by a large indentation. The description of the simulation error shown in Equation 4.8 is then rewritten in the discrete form as

$$S_{err} = \frac{\sqrt{\sum_i \left(F_s(i) - F_e(i) \right)^2}}{\sqrt{\sum_i \left(F_e(i) \right)^2}} \quad (4.10)$$

The best soft tissue parameters can be optimised by minimising this error factor using an iterative procedure such as the *fminsearch* function provided by MATLAB®. Typical experimental and simulated data can be seen in Figure 4.22. In addition, Ling et al. (2007a) used a genetic algorithm to optimise the searching process where Poisson's ratio could be also found from the optimisation process.

When a spherical indenter is used for the test and the tissue thickness is much larger than the indenter radius, Young's modulus can be calculated using the following simplified equation assuming a single solid state of the indented soft tissue (Yang et al. 2007):

$$E = \frac{3(1-v^2)}{4R^{1/2}} \frac{F}{d^{3/2}} \quad (4.11)$$

where
   F is the indentation force
   R is the indenter radius
   d is the indentation depth

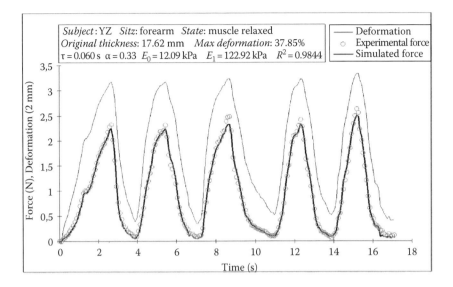

FIGURE 4.22 Typical indentation curves, which include the experimental indentation, experimental force and the simulated force based on indentation history using a quasilinear viscoelastic (QLV) model.

If the indenter's radius is comparable to the tissue thickness, a simulation method such as finite element analysis can be adopted for solving the problem (Lee et al. 2011).

When the contact surface is in a rectangular shape, for example most of the ultrasound probes for 2D imaging, and assuming the tissue with a single solid phase and an infinitesimal deformation, Young's modulus of the tested soft tissue can be calculated as (Alblas and Kuipers 1970; Han et al. 2003)

$$E = 2\left(1 - v^2\right)\kappa\left(v, \frac{s}{h}\right)\frac{F}{dl} \tag{4.12}$$

where

    F is the indentation force
    d is the indentation deformation
    v is the Poisson's ratio
    κ is a scaling factor related to Poisson's ratio v
    s is the length of the short side of the probe
    l is the length of long side of the probe
    κ is a scaling factor which depends on the Poisson's ratio and the aspect ratio s/h

The value of κ can be obtained by an approximate mathematical formulation (Alblas and Kuipers 1970) or numerically by finite element analysis.

In the case of non-contact indentation methods, to the best of our knowledge, no theoretical analysis has been found in the literature. Lu et al. (2012) adopted a one-way fluid–solid coupling analysis to simulate the interaction of water jet and soft tissue using the finite element method and also proposed a revised scaling factor κ for the extraction of Young's modulus from the water-jet indentation test:

$$E = \frac{\left(1 - v^2\right)}{2a\kappa\left(v, a/h, d/h\right)} \cdot \frac{F}{w} \tag{4.13}$$

where the scaling factor κ is dependent on Poisson's ratio v, the aspect ratio a/h and the indentation/initial thickness d/h. The interaction nonlinearity coming from the finite deformation was demonstrated by varying κ with respect to the deformation amplitude. Alternatively for simplicity, a direct stiffness coefficient K, which is similar to Young's modulus in a compression test, can also be used for a good representation of the elasticity of the tested object (Lu et al. 2005, 2007):

$$K = \frac{F/\left(\pi a^2\right)}{d/h} \tag{4.14}$$

where the same contact area (circular area of radius a) is assumed for the interaction between the water jet and soft tissue. This parameter approaches Young's modulus when the initial thickness of the soft tissue is comparable to the water jet radius. More theoretical studies of the fluid jet–induced non-contact indentation are needed in the future for a thorough analysis of the interaction between the fluid medium and soft tissue.

4.2.3 Reliability and Accuracy Analysis of Measurement

When applied to clinical test for tissues, a novel indentation test's reliability and accuracy should be considered. When tested at the same site, the repeatability of the ultrasound indentation test was high (Zheng et al. 1999). With respect to ultrasound indentation, the perpendicularity between the ultrasound probe and the soft tissue surface should be particularly considered, as it is designed for hand-held test *in vivo*. Zheng et al. (1999) studied the effect of the indentation angle on the measurement results for soft tissues *in vitro* and *in vivo* (Figure 4.7). Young's modulus obtained with approximately 5°–10° of misalignment was consistently smaller than that obtained with maximum strength signal alignment. The effects became smaller when the thickness became larger, and it was almost negligible when the thickness was larger than twice the indenter diameter. In addition to Young's modulus, the ultrasound signal is also significantly affected by the misalignment. When the ultrasound beam is perpendicular to the detected objective surface, the signal amplitude will reach a maximum value which is a good reflection of the alignment. Therefore, the amplitude of the reflected signal at the interface can be used as an indicator to judge the perpendicularity between the ultrasound transducer and the contact surface to reduce its effect on the measurement of tissue thickness and elasticity. Another parameter affecting the reliability of the measurement is the indentation rate. Zheng et al. studied the effect of the indentation rate on the elasticity measurement (Zheng 1997; Zheng et al. 1999b). Manual indentation, which was controlled in the range 0.75–7.5 mm/s for the indentation test of forearm soft tissues, was used. The results showed that the effective Young's modulus was not sensitive to the indentation rate in the tested range and the standard deviation for different rates was less than 10%. In addition, the principle of ultrasonic measurement of tissue thickness is based on a hypothesised constant velocity of the ultrasound in soft tissues. The sound speed in soft tissues should be measured precisely for an accurate measurement of the tissue thickness and deformation. Therefore, the accuracy of ultrasound speed will affect the precision of stiffness measurement. The best way to solve this is to calibrate the speed of sound in the tested tissue, whereby the thickness and deformation measurement accuracy can be improved. Although the measurement of sound speed has been successfully developed for bone tissue in order to assess its health index, no simple method has been developed for such measurements in soft tissues *in vivo*. An easy and convenient method should be developed in the future for this purpose. When 2D ultrasound is used, the measurement of thickness will also be affected by the algorithms for the detection of the objective surface. Kawchuk and Elliot (1998) compared the difference of displacements measured by 2D ultrasound and by a standard machine. In their experiment, a 5 MHz ultrasound probe was used, and the results showed that, even after speed calibration, the detection error was related to the detection algorithm. The average error was ~14%–22%. It should be noted that the accuracy of ultrasonic measurement is mainly decided by the characteristics of the ultrasound transducer, including its frequency, focusing and signal format (raw RF/demodulated amplitude/image signals). Therefore, the measurement accuracy and reliability need to be tested individually for each specific measurement system, as the information cannot be simply adopted from other systems.

4.2.4 Related Measurement of Multiple Tissue Properties

Associated with novel indentation techniques, multiple tissue properties can also be measured for a better characterisation of soft tissues. For example, with the ultrasound indentation, possible parameters that can be measured include the speed of sound, surface reflection, attenuation and scattering coefficient. Taking the articular cartilage as an example, Suh et al. (2001) proposed an ultrasound indention method that could measure the speed of sound in the articular cartilage. Through this method, it was demonstrated that the speed of sound in cartilage after degeneration induced by enzymatic digestion of proteoglycans was significantly smaller than that before the degeneration (1598 ± 28 vs. 1735 ± 35 m/s). However, when the speed was measured in naturally generated cartilage, it was slightly smaller than that of normal cartilage and the difference was not very obvious (Myers et al. 1995; Toyras et al. 2003). Research also showed that the speed is affected by the magnitude of compression and the compression rate (Patil 2005; Ling et al. 2007b; Nieminen et al. 2007; Kiviranta et al. 2009). Therefore, whether the ultrasound speed can be used as an objective indicator of the degeneration status still needs further investigation (Nieminen et al. 2009). Another commonly used parameter for the characterisation of cartilage degeneration is the surface acoustic reflection coefficient, which can be defined both in the time domain and frequency domain. A reference signal reflected by a perfect reflector placed at the same distance as the tissue sample should be obtained for calibrating the real ultrasound signal to get rid of the system effects. After calibration, a system-independent reflection coefficient can be obtained for the tissue characterisation (Laasanen et al. 2002; Wang et al. 2010b). Using a model of enzymatically induced degeneration of articular cartilage, it has been demonstrated that the surface reflection would significantly decrease after the collagen network was broken down in the cartilage (Toyras et al. 1999; Laasanen et al. 2002; Lu et al. 2009a; Wang et al. 2010b). When this method was applied to the naturally degenerated cartilage on animal or human samples, it also showed that the surface reflection decreased with the increased severity of degeneration (Saarakkala et al. 2003; Kiviranta et al. 2008). Therefore, the surface reflection coefficient is thought to have a high potential as an objective indicator of the cartilage degeneration in clinical studies. Some studies also adopted ultrasound attenuation or scattering to investigate its potential in characterising cartilage degeneration. However, not so many studies exist in this field and the current results do not provide much support for this direction (Nieminen et al. 2009).

Ultrasound can also be used to measure the morphological properties of the cartilage. For example, in articular cartilage, ultrasound can be used to assess the surface roughness of the cartilage. Saarakkala et al. (2004) proposed an ultrasound roughness index (URI) to study the degeneration of the cartilage. As there will be some fibrillation on the surface of the cartilage with delicate degeneration, the surface roughness will increase. Therefore, the study of cartilage morphology has the potential to detect the early change of cartilage associated with degeneration. An endoscopic ultrasound probe with imaging function has the potential to assess the morphological change of the cartilage, such as intravascular ultrasound, for the assessment of cartilage degeneration (Huang and Zheng 2009; Viren et al. 2009, 2010, 2011). An arthroscopy-based ultrasound probe has been developed by the authors' group. Miniature ultrasound imaging was adopted in the arthroscopic channel where the ultrasound indentation could also be performed for the morphological, acoustical and mechanical assessment of articular cartilage and its degeneration (Huang and Zheng 2013). Correspondingly, if optical imaging methods are adopted in the mechanical test, such as the OCT-based air-jet indentation test, the related optical properties including refractive index, optical surface reflection, optical scattering coefficient and other morphological parameters can also be obtained from the soft tissue for the assessment of tissue physiology and pathology (Saarakkala et al. 2009; Wang et al. 2010c; Huang et al. 2011a).

4.3 NANOINDENTATION

When the indentation scales are on a nanometer resolution, such as a nanometer scale of deformation or alternatively a nano/micro scale of indentation force, it is called a nanoindentation, which has become a very important method for the assessment of materials (Cook 2010). Why is it necessary to develop the nanoindentation technique for the measurement of mechanical properties of biological tissues? The diversity of biological tissues has made their mechanical properties very heterogeneous, and therefore it is necessary to study the different parts of an organ or a tissue to analyse the distribution of their mechanical properties. Furthermore, some tissues are very small, for example the articular cartilage of small animals such as mouse. The total thickness is very small, and only nanoindentation is the most suitable technique for obtaining its mechanical properties. When nanoindentation is introduced to the field of testing biological soft tissues, it will encounter some new problems, as it was originally developed specifically for testing industrial materials. The factors concerned include a much smaller stiffness, nonlinear stress/strain relationship and the viscosity and the heterogeneity of the biological soft tissues compared to industrial materials such as metals. Among them, effects of the main two issues, i.e. the much smaller stiffness and time-related viscosity, are briefly discussed in the following.

Compared to a high elastic modulus on the order of gigapascals for a metal, the effective modulus of soft tissues is in the scale of kilopascals to megapascals. Therefore, when the same indentation is used for the test, the force response in soft tissues will be much smaller than that on metals, and thus it requires a very high precision for force measurements. As the indentation force is proportional to the contact area, one way to solve this is to increase the indenter size (the radius for a spherical/cylindrical indenter or the tip angle for a pyramidal indenter) so that the force can be enlarged correspondingly. However, when the size of the indenter is increased, the spatial resolution of the nanoindentation will be compromised. As indentation of soft tissues will not leave irreversible deformation to the tested object, the distance between two measurement points can be reduced a little to improve the spatial resolution of measurement when a larger indenter is used with a small amplitude of indentation. Another issue associated with the nanoindentation of soft tissue with a small elasticity modulus is the detection of contact. In traditional nanoindentation, a small indentation force or an abrupt change of the indentation/deformation (stress/strain) is used for the detection of contact of the indenter with the material (Fischer-Cripps 2006). However, as a small indentation force is induced in the test of soft tissues, the algorithms with automatic detection on the testing of traditional materials may no longer be appropriate. An alternative way to solve this problem of contact point detection is to find the point manually and then move the specimen surface by a small distance before starting the real nanoindentation test (Kaufman et al. 2008; Kaufman and Klapperich 2009).

A problem with nanoindentation is the surface adhesion issue. This issue is not important in macroscopic indentation; however, it will significantly affect the results in the nanoindentation during unloading, as it will significantly increase the contact area. Related compensation methods have been proposed for reducing its effects on measurement (Carrillo et al. 2005; Ebenstein and Wahl 2006).

The traditional nanoindentation adopts the classical Oliver–Pharr method to analyse the data obtained from the nanoindentation test (Oliver and Pharr 1992, 2004):

$$E = \frac{\sqrt{\pi}\left(1-v^2\right)}{2\sqrt{A}} \frac{dF}{d\delta} \tag{4.15}$$

where

   A is the contact area when the maximum indentation force is reached
   v is Poisson's ratio
   F is the indentation force
   δ is the indentation deformation
   $dF/d\delta$ is the slope obtained from the initial unloading phase of the
      curve (Figure 4.23a)

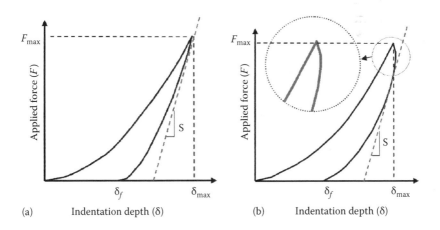

FIGURE 4.23 Typical curve of nanoindentation test where the indentation slope S is calculated for the extraction of Young's modulus of the tested material. (a) Test on a solid material. (b) Test on a soft tissue with viscosity.

When using Equation 4.15, pure elasticity is assumed for the tested object and a triangular force/time curve is utilised in the test. When biological soft tissues are tested, creep phenomenon exists and this may cause a 'nose' phenomenon (Figure 4.23b) when a triangular load function is used (Ebenstein and Pruitt 2006). The nose effect will cause the calculation of the indentation slope to be inaccurate, and a new load function such as the trapezoidal function and correction algorithms were proposed to solve this problem (Feng and Ngan 2002; Cheng et al. 2005). The principle is to hold the load peak long enough to let the material reach the equilibrium state without any further creep before the unloading starts (Briscoe et al. 1998). On the other hand, because of the viscoelasticity of biological soft tissues, the whole Oliver–Pharr algorithm can be abandoned for the measurement of Young's modulus on soft tissues. Instead of calculating the parameter in the unloading phase, it would be better to obtain the tissue's viscoelastic parameters in the loading phase (Oyen 2005), just as was done in the macroscopic indentation. In addition to the quasi-static test, a dynamic test such as cyclic indentation can also be performed to study the change of material properties along with the testing frequency (Franke et al. 2011). With respect to the difference, difficulty and characterisation of nanoindentation between soft tissues and traditional industrial materials, the readers can refer to some of previous excellent review papers and books for more information (Ebenstein and Pruitt 2006; Franke et al. 2008; Lin and Horkay 2008; Oyen 2011a,b).

4.4 FUTURE DEVELOPMENT OF INDENTATION

In this chapter, several novel indentation techniques, mainly ultrasound indentation, optical indentation, fluid jet indentation and nanoindentation, are introduced for the measurement of elasticity in soft tissues. Some characteristics of these novel indentation techniques include the portability, measurement of multiple tissue properties and measurement of elasticity of a very small specimen. Although these novel indentation techniques have been applied in different situations, they have not become popular clinical tools so far because of various limitations and disadvantages encountered in practical applications. To apply these techniques in a variety of clinical fields requires broad collaborations of R&D engineers and clinicians with input of efforts from both sides. As engineers, the authors propose some future directions in the following aspects for a better development of this field: (1) develop specific instruments for the measurement of elasticity in certain body parts, for example the development of miniaturised devices for the arthroscopic measurement of intra-joint tissue properties; (2) further improve the measurement precision and reliability; (3) improve ease of operation of devices and provide user-friendly interface; (4) calculate the intrinsic tissue elastic parameters based on test models and (5) set up brief instructions for the operations of the devices by the clinical practitioners, so that a broad range of clinical trials can be performed.

CHAPTER 5

Suction Measurement of Tissue Elasticity *in Vivo*

5.1 INTRODUCTION

In the last two chapters, the indentation test has been thoroughly discussed for the purpose of the measurement of soft tissue elasticity. The advantage of indentation is obvious: i.e. the tissue need not be dissected so that the test can be performed in a very convenient way *in vivo*. Therefore, it has become a very popular method to measure the biomechanical properties of soft tissue. However, one disadvantage of the indentation test is that it also needs a firm foundation for the tested soft tissue so that the test can be conducted in a reliable and accurate way. Otherwise, the motion of the soft tissues caused by heartbeat, respiration or body motion during the test will become a very important source of error and make the test results unsuitable for analysis. On the other hand, based on our own experience, when the tissue is too soft, it is not so easy to conduct a macroscopic indentation test because the force response from the tested tissue is small and measurement of the small force during the test *in vivo* will not be accurate. Again, in our experience, when Young's modulus of the soft tissue is less than 10 kPa, testing by manual indentation will become quite difficult. Another problem encountered in practical indentation measurement of soft tissue elasticity is that, when it is necessary to measure the elasticity characteristics of the superficial tissue layer, a smaller indenter should be used. In the extreme case, nanoindentation is the potential choice. However, when the indenter becomes smaller and smaller, a practical issue that is similar to the test on very soft tissue is how to measure the small force response reliably and accurately. In addition, because of the layered structure of tissues, a small indenter can sense the elasticity of only the superficial tissue layer. Therefore, it is necessary to develop other measurement methods that have the potential to solve the problems mentioned before.

The suction (also called aspiration) test, developed to measure the mechanical properties of soft tissues *in vivo*, has more potential to solve the aforementioned difficulties. In the suction test, a tube with a central air chamber is placed on the surface of the soft tissue (Figure 5.1). Then a negative pressure is created in the chamber so that part of the tissue is lifted into the tube. A pressure sensor is used to record the negative pressure, and other methods such as optical sensing can be used to record the height of the lifted tissue. Parameters such as the indentation force and deformation in the indentation test are used to extract the mechanical properties from the tested soft tissue. In principle, suction can be regarded as the opposite of indentation; whereas indentation compresses the tissue

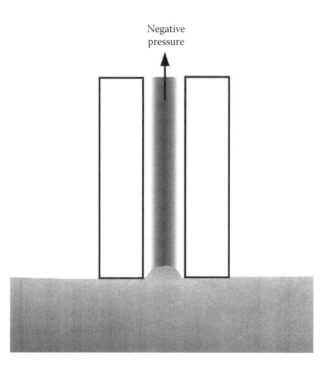

FIGURE 5.1 Suction probe in use. It has a small hole at the centre of contact area to provide a negative pressure to aspirate the tested tissue.

locally, suction extends the tissue in a local way, so both tests can be conducted on living tissues. There are similarities and differences between the two tests, some of which are the following: Most part of the tissue endures compression in indentation but extension in suction. The deeper layer of the tissue serves as a support for indentation, while the plantar surface of the suction tube serves as the support for suction. The force–deformation relationship is significantly dependent on the ratio between the tissue thickness and the indenter size in indentation, while this is also the case for the suction where the behaviour is also dependent on the ratio of the dimension between the tissue thickness and the air chamber size. In suction, the tissue can be attracted and attached to the bottom surface of the tube so that the relative motion of the organ would not have too much effect on the test results, which is not the situation in the indentation test. On the other side, the suction strength is measured in terms of air pressure, not as the direct force as it is in the indentation test. As the pressure that can be measured has a relatively small value (in the range of millibars), the suction method can easily measure a relatively small value of Young's modulus in a soft-tissue test. Zheng et al. (2014) reported a probe that could perform both suction and air-jet indentation, which can potentially provide complementary advantages of the two tests (Zheng et al. 2014).

The suction test has been applied to the assessment of a variety of soft tissues, including the skin, liver, brain, kidney, cervix and the vocal fold. Skin was probably the first tissue where suction was applied for the measurement of its mechanical properties in physiological situation or pathologies. Grahame adopted the suction method to study the spatial variation of the mechanical properties of skin and their change with ageing, sex and pregnancy (Grahame and Holt 1969; Grahame 1970). In order to reduce the effect of natural tension of the skin during the test, Cook et al. developed a pre-tensioning device to test the mechanical behaviour of the skin by counter-balancing the natural skin tension (Alexander and Cook 1977; Cook et al. 1977). Later, some suction devices for skin, which included the Dermaflex® (Gniadecka and Serup 2006), Cutometer® (Elsner et al. 1990; O'goshi 2006) and DermaLab (Grove et al. 2006), became commercially available in the market. Today, suction devices such as Cutometer have become standard clinical tools for the measurement of the mechanical properties of the skin in order to assess skin physiology and pathology and to evaluate the effectiveness of treatment regimens (Enomoto et al. 1996; Romano et al. 2003; Dobrev 2005; Ryu et al. 2008). The differences between different suction devices are shape, profile, chamber size and method of attachment between the probe and skin surface. It should be noted that the method to measure the displacement of the skin also changed along with the development of the devices. Initial studies used a linear variable differential transformer (LVDT) installed in the inner chamber or the change in capacitance between the skin and the electrode to measure the height of the aspirated tissue (Cook et al. 1977; Gniadecka and Serup 2006), while later devices utilised the optical method or ultrasound imaging to measure this parameter (Diridollou et al. 2000; O'goshi 2006) with improved spatial resolution or operational convenience. As the skin is a layered structure, the aperture size of the suction device will greatly affect the test results. When the aperture size is small, the top layers of the skin including the epidermis and dermis mainly contribute to the test behaviour. However, when a large aperture is used, the subcutaneous tissue may also contribute significantly to the test results. Hendricks et al. used a two-layer finite element model to study the mechanical properties of skin using different aperture sizes and found that the ratio of the two layers can be as large as 1000, which might lead to convergence problems in the analysis (Hendriks et al. 2006). The suction test in combination with finite element analysis has become a common and powerful tool for the study of the mechanical properties of the skin (Hendriks et al. 2003; Barbarino et al. 2011). In the last decade, the suction test has also been introduced for the study of other soft tissues, such as the uterus, cervix, brain, liver and vocal folds. A research group from ETH, Zurich, Switzerland, has developed a suction device in which a mirror is placed on the side of the tube so that the profile of the attracted tissue under suction could be clearly observed by a camera for analysis (Vuskovic 2001). This device was specially designed to measure the mechanical properties of internal organs. *In vivo* experiments showed that with this device it was possible to study the pathologies of soft tissues, such as cervical insufficiency during gestation (Mazza et al. 2006; Bauer et al. 2009) and liver diseases (Nava et al. 2004; Mazza et al. 2007). Schiavone et al. also adopted a similar device for the measurement of elasticity of brain tissue (Schiavone et al. 2009). A general difficulty in adopting the suction device developed by Vuskovic (2001) for brain study was that it was too big and heavy to fit in the limited space and soft tissue during brain surgery. Recently, a suction device was reported using fibre-based optical coherence tomography imaging (Zheng et al. 2014), which provided the possibility of a very small suction probe. The ease of sterilisation is another important issue that needs to be considered for a device operated during the surgery of internal organs.

Therefore, a light suction device was developed by using plastic material so that it was possible to make the operation of the suction device quite easy without manual operation during the test. Later, a miniaturised metallic probe was developed, in which a digital camera was also embedded so it was possible to be applied for *in vivo* measurement of brain tissues (Schiavone et al. 2010). Furthermore, a library-based optimisation algorithm made the extraction of the mechanical parameters of the measured tissue available immediately after the operation on site, so that it had the potential to be developed as a clinical tool for the objective assessment of brain tissues.

Besides macroscopic operation, the suction test has also been adopted in mechanobiology for the assessment of mechanical properties of cells (Figure 5.2) (Hochmuth 2000; Lim et al. 2006; Addae-Mensah and Wikswo 2008). In contrast to a macroscopic suction test, a microscopic imaging system is necessary to be incorporated in the system set-up to observe the response of the tiny cell. The cell suction test can be traced back to the 1950s, when a simple suction device was used to assess the mechanical properties of the cell membrane (Mitchison and Swann 1954). A water reservoir, which could be moved in the vertical direction, was used to adjust the suction pressure in this study, and certainly this could be improved by using modern pumping devices. One big difference of the suction test for the cell compared to that for macroscopic soft tissues is that the cell is divided into two different types, with one behaving like a liquid drop such as the neutrophil and the other like a solid such as the chondrocyte (Hochmuth 2000). For the cell suction test, the cell is aspirated into the pipette until a hemisphere is formed in the pipette. After a

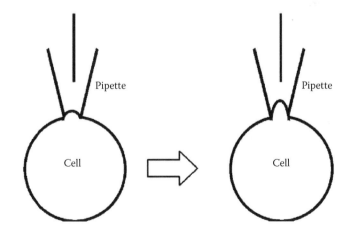

FIGURE 5.2   Illustration of suction test for a cell using a micropipette.

critical point, further increase in suction pressure leads to a liquid-like cell that will be totally aspirated into the pipette, while a solid-like cell will be further aspirated into the pipette until reaching a new equilibrium state. Suction could also be used to study the cell adhesion by using a pair of suction devices (Shao et al. 2004). Suction combined with fluorescence imaging could be used to observe the strain distribution in the cell membrane under deformation (Discher et al. 1994).

An important issue in the suction test of soft tissues is how to analyse the mechanical properties of the tissue from the collected data, which is presented in detail in the following section. Simple phenomenological analyses are introduced first, and then some methods for the extraction of the intrinsic parameters are presented. Finally, some special analyses for the suction test on the cell are presented, as this test has become more and more popular for biomechanical studies of cells.

5.2 ANALYSIS OF SUCTION TEST

5.2.1 Macroscopic Suction

From the suction test, usually two parameters, i.e. the negative pressure P and the displacement d of the apex of the absorbed tissue, as shown in Figure 5.3, are measured as a function of time. As the inverse problem is quite complex, a simple way to analyse the data is to directly use the ratio of the two parameters P and d as an index of the tissue properties. A typical suction test is conducted by an abrupt increase of the negative pressure and then keeping for a while until reaching a final steady status. A typical curve of the deformation with time is shown in Figure 5.4 for the Cutometer, and various parameters can be defined from this curve as shown in Table 5.1 (O'goshi 2006). It should be noted that these defined parameters can be compared between different subjects, or in a longitudinal study only when the same level of air pressure is used for the test. Among all the parameters, Ue can be regarded as a good representation of the skin stiffness and Uv/Ue as characteristic of the skin viscosity. It should be admitted that although quite a number of parameters have been defined for the tested tissue, it is not easy to explain the exact difference between some of these parameters and relate them to the physiology or mechanics. Ideally, intrinsic material parameters should be used for tissue assessment, as they are independent of the measurement method, suction chamber, vacuum pressure, rate of pressure change, etc. (Zheng and Huang 2008). In this way, the comparison between different tests, particularly

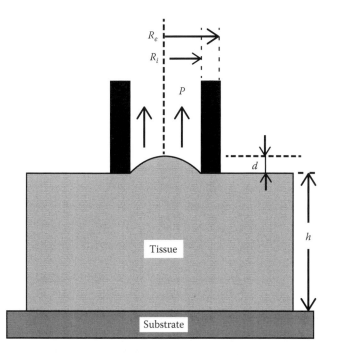

FIGURE 5.3   Schematic showing the suction test of a soft tissue.

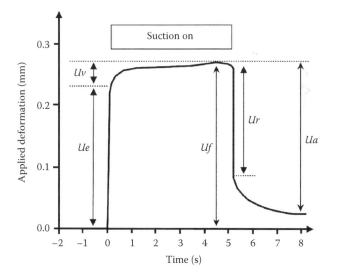

FIGURE 5.4   Typical suction curve for the analysis of tissue properties. (Modified from O'goshi, K., Chapter 66: Suction chamber method for measurement of skin mechanics: The Cutometer®, in: Serup, J. and Jemec, G., eds., *Handbook of Non-Invasive Methods and the Skin*, CRC, Boca Raton, FL, 2006.)

TABLE 5.1   Common Parameters Adopted in the Suction Test Using a Cutometer

Parameters	Interpretation
Ue	Immediate deformation: extensibility
Uv	Viscoelastic deformation: plasticity
Uf	Final deformation: distensibility
Ur	Immediate retraction
Ua	Final retraction after removal of vacuum
Ua/Uf	Gross elasticity
Ur/Ue	Pure elasticity
Ur/Uf	Biological elasticity
Uv/Ue	Ratio between delayed and immediate deformation

Source: O'goshi, K., Chapter 66: Suction chamber method for measurement of skin mechanics: The Cutometer®, in: Serup, J. and Jemec, G., eds., *Handbook of Non-Invasive Methods and the Skin*, CRC, Boca Raton, FL, 2006.

conducted by different measurement systems and by different groups, will be easier and more meaningful.

It is more appropriate to include the suction pressure in the calculation of the mechanical parameters. When a cyclic suction test with constant vacuum pressure $p_{\min}$ is used, three parameters, i.e. a 'stiffness' parameter k (mbar/mm), a dimensionless 'creep' parameter δ and a dimensionless 'softening' parameter γ, can be defined as follows (Figure 5.5) (Mazza et al. 2006):

$$k = \frac{p_{\min}}{d_1} \tag{5.1}$$

$$\delta = \frac{d_1}{d_0} \tag{5.2}$$

$$\gamma = \frac{d_4 - d_1}{d_1} \tag{5.3}$$

Of these parameters, k is similar to the force/deformation parameter in the indentation test, which indicates how the tissue is deformed under unit pressure, and δ and γ are parameters indicating the extent of creep in a single cycle and after four cycles of suction, respectively. It should be pointed out, again, that these parameters are comparable only under certain assumptions of tests, such as using the same suction device and the same pressure history. In order to extract intrinsic material properties from the suction test, further analysis is necessary.

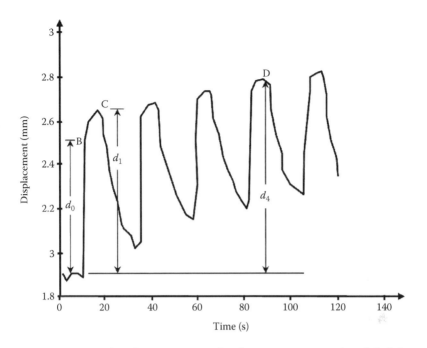

FIGURE 5.5 Typical cyclic suction test for the uterine cervix. (Modified from Mazza, E. et al., *Med. Image Anal.,* 10, 125, 2006.)

When the tested material is assumed to be homogeneous, isotropic and incompressible, and the contact between the tissue and pipette is smooth without friction, the material's Young's modulus can be measured from the following equation (Figure 5.3) (Theret et al. 1988):

$$E = \frac{3\phi(\eta)}{2\pi} \frac{P}{\delta} \tag{5.4}$$

where
$\delta = d/R_i$ is the apex displacement with respect to the pipette radius
P is the negative pressure
$\eta = (R_e - R_i)/R_i$ is a shape factor of the pipette, defined as the ratio of pipette thickness to its inner radius
ϕ is a function of pipette shape factor η

When $0.4 \leq \eta \leq 0.6$, a mean value of $\phi = 2.066$ can be used. In this case, Equation 5.4 can be rewritten as

$$E = 0.986 \frac{P}{\delta} \tag{5.5}$$

The coefficient 0.986 is very similar to that in another two studies (Aoki et al. 1997; Alexopoulos et al. 2003). However, as to the indentation test, the limited thickness of the tested tissue, h, with respect to the pipette radius R_i should be considered in practical situations, and therefore there should be some correction for the equation when h is comparable to R_i. When the thickness of the tissue is considered and the material has Poisson's ratio v, a correction factor $\alpha(v, \xi)$ is used for the calculation (Boudou et al. 2006):

$$E = \alpha(v, \xi) \frac{3\phi(\eta)}{2\pi} \frac{P}{\delta} \tag{5.6}$$

where $\xi = h/R_i$ is a dimensionless parameter representing the tissue thickness. The correction factor $\alpha(v, \xi)$ can be approximated by the following equation based on computational simulation results:

$$\alpha(v, \xi) = H(v) \left(\frac{\xi^{p(v)}}{K(v) + \xi^{p(v)}} \right)^n \tag{5.7}$$

where $n = 1$ for an adherent sample and $n = -1$ for a non-adherent sample with respect to the substrate. The various coefficients are defined as polynomials of Poisson's ratio v:

$$H(v) = H_2 v^2 + H_1 v + H_0 \tag{5.8}$$

$$K(v) = K_3 v^3 + K_2 v^2 + K_1 v + K_0 \tag{5.9}$$

$$p(v) = p_4 v^4 + p_3 v^3 + p_2 v^2 + p_1 v + p_0 \tag{5.10}$$

The various coefficients of these polynomials can be found in the paper by Boudou et al. (2006). Furthermore, the analytical solution of $\alpha(v, \xi)$ can be used as a novel method to extract Poisson's ratio from the suction test. The method is to perform two suction tests on samples with different ξ values, denoted as ξ_1 and ξ_2. Then the ratio of the two correction factors for an adherent sample can be written as

$$R(v, \xi_1, \xi_2) = \frac{\alpha(v, \xi_1)}{\alpha(v, \xi_2)} = \left(\frac{\xi_1}{\xi_2} \right)^{p(v)} \frac{K(v) + \xi_2^{p(v)}}{K(v) + \xi_1^{p(v)}} = \frac{P_2 \delta_1}{P_1 \delta_2} \tag{5.11}$$

By solving Equation 5.11 in the interval $0 \leq \nu \leq 0.5$, Poisson's ratio can be obtained as a material property. This idea of obtaining Poisson's ratio is very similar to the method used by indentation with two different sized indenters (Jin and Lewis 2004; Choi and Zheng 2005), as introduced in Section 3.2.1. After obtaining Poisson's ratio, Young's modulus of the soft tissue can be further calculated using Equation 5.6. An alternative way to solve this problem is to use the graphic method (Boudou et al. 2006). When ξ_2 is chosen to be infinitely large, and with $\xi_1 = \xi$, the ratio can be rewritten as

$$R\left(\nu, \xi, \infty\right) = \frac{\xi^{p(\nu)}}{K\left(\nu\right) + \xi^{p(\nu)}} = \frac{P_2 \delta_1}{P_1 \delta_2} \tag{5.12}$$

The right side of Equation 5.12 can be used to calculate the value of R. A graph relates the value of R with respect to ξ for different ν values. After obtaining the values of R and ξ, a specific point M is defined, and the corresponding ν can be found for which its curve passes through the point M. After getting the value of ν, the corresponding value of α can be obtained so that Young's modulus of the tested tissue can be finally calculated by using Equation 5.6.

More complex models can be considered for the mechanical behaviour of soft tissues in suction (Kauer et al. 2002; Hendriks et al. 2003, 2006; Boudou et al. 2006; Barbarino et al. 2011). When such models are used, the model parameters cannot be easily extracted from a theoretical analysis of the suction problem. In these circumstances, finite element analysis is a more appropriate method to solve the inverse problem. When the mechanical model of the tissue is defined, the whole process of analysis is not much different from that of the indentation analysis as seen in Chapter 9, where the finite element analysis method is discussed.

5.2.2 Microscopic Suction

The microscopic suction method was developed to measure the elastic properties of small tissues, mainly the cell. It is hypothesised that the change in mechanical properties of the cell is associated with tissue diseases, and therefore it is important to study the relationship between cell mechanics and tissue diseases. Compared to the macroscopic test, the main challenge of microscopic suction comes from the small scales of the measured parameters. For example, the main units for the distance, force and pressure are in the order of μm, pN–nN and Pa–kPa in microscopic suction, respectively, whereas these are in the order of mm–cm, mN–N and kPa–100 kPa in macroscopic suction. Because of the small size, all the operations of the suction test of cells are conducted under the guidance of a microscope. The analysis of a suction test on a cell membrane can be divided into two situations depending on whether the cell behaves more like a liquid or a solid. Considering the equilibrium state, the cortical tension is an important parameter to characterise the liquid-like cell such as neutrophil or red blood cells (Evans and Yeung 1989; Hochmuth 2000). Considering the cell as a sphere, the cortical tension (T_c) can be measured from the critical point where the aspirated tissue height is equal to the pipette radius:

$$T_c = \frac{P_c}{2\left(\dfrac{1}{R_i} - \dfrac{1}{R_c}\right)} \tag{5.13}$$

where
   P_c is the critical aspiration pressure
   R_i is the pipette radius
   R_c is the radius of the cell outside the pipette

When the cell is not spherical but like a disc, the membrane shear modulus is the most important parameter for analysing the cell properties, which can be approximately calculated using the following equation (Chien et al. 1978):

$$\mu = 0.41 R_i^2 \frac{\Delta P}{\Delta d}, \quad \left(\text{where } \frac{d}{R_i} > 1\right) \tag{5.14}$$

where
   $\Delta P/\Delta d$ is the regression of pressure versus tissue height
   R_i is the pipette radius

Please be noted that the shear modulus is defined in unit of force/deformation in this case.

Through experimental studies, it was found that the shear modulus of the red cell (6–9 pN/μm) was about one-fourth of the cortical tension of the neutrophil (~35 pN/μm). For a solid-like cell, the elastic modulus can also be calculated using Equation 5.4. With this elastic modulus, it is also possible to calculate the cortical tension at the critical point where the aspirated tissue height is equal to the pipette radius. The calculated cortical tension for chondrocyte was ~2200 pN/μm, which was significantly larger than that of the soft neutrophil (Hochmuth 2000).

5.3 SUMMARY

The suction test is a viable method for the measurement of soft tissue elasticity in situations where the traditional indentation may not be applicable. These situations include the test on soft tissues with no firm substrate or whose movement is large in physiology, as well as tests on soft tissues with a relatively small Young's modulus such as the liver and the brain. With simple assumptions of the tissue properties, Young's modulus of the tissue can be calculated with the measured pressure and deformation data using a simple equation. However, when a more complex mechanical modelling of the tissue behaviours is necessary, an inverse problem should be solved to obtain the model parameters, usually through a numerical computational method such as finite element analysis. The suction method is a very useful method to study the mechanical properties of soft tissues, particularly the micropipette aspiration on cells because cellular mechanobiology has been recognised as an important issue in disease pathophysiology (Ingber 2003; Wang and Thampatty 2006). However, just like the indentation test, the suction test has some limitations. First, it is easily affected by the boundary conditions of the test, including the tissue geometries and contact conditions. Second, it can only measure the local properties of the superficial layer of the tissue in contact with the suction device. It is not effective in measuring the elasticity of deep soft tissue noninvasively, which is necessary for broad clinical applications *in vivo*. These problems can be partially solved by the new methods that have been developed in the last decades, which include the shear wave propagation method. Some of these recently developed methods are presented in the next chapter.

CHAPTER 6

Indirect Methods for Soft Tissue Elasticity Measurement *in Vivo*

6.1 INTRODUCTION

Test methods used in traditional material testing have been adopted to measure the mechanical properties of soft tissues directly. In these methods, certain stimulation to tissue, such as the force in compression or indentation, is used to induce some perturbation to the tissue structure. Based on the relationship between the stress and the strain, the stiffness of the tissue can be calculated, as in the case of traditional compression, extension or indentation tests. However, these methods are quite complex when the test is conducted on tissues with irregular geometry or when there is large spatial inhomogeneity of the tissue elasticity. The problem then belongs to an inverse problem as seen in the development of elastography, where the strain elastogram is used to reconstruct the modulus mapping of the tested tissue (Kallel and Bertrand 1996). Usually, these questions do not have theoretical solutions because of the very complicated geometry and boundary conditions with varied spatial distribution of the tissue properties. Finite element analysis (FEA) is a viable computational method to solve this problem. However, the computational cost for FEA is very high because of a large number of unknown variables and, sometimes, because of convergence problems during computing. To effectively solve this problem, indirect methods, by resorting to the measurement of other parameters, are worthy of investigation.

In this chapter, some indirect methods for the measurement of tissue elasticity are introduced. As the name suggests, indirect methods enable the measurement of tissue elasticity by the use of some probing factors that reflect the tissues' biomechanical properties. These methods can be classified into two categories: (1) detection of the resonant frequency change of a probe when contacting tissues with different elasticity (Omata and Terunuma 1992), and (2) detection of shear wave propagation in tissues, including holographic wave pattern monitoring on the tissue surface (Avenhaus et al. 2001) and propagation speed measurement in tissue to indicate its elasticity (Sarvazyan et al. 1998). Recently, methods using the shear wave propagation speed have developed rapidly, and a number of products for different purposes are now available. Details of each method are introduced in the following sections.

6.2 RESONANT FREQUENCY SHIFT MEASUREMENT

Tissue elasticity can be measured using a tactile sensor that detects the shift of the resonance frequency after the sensor contacts the tissue (Omata and Terunuma 1992; Omata et al. 2004). The sensor has its own resonant frequency when there is no contact with the tissue (Figure 6.1). The loading acoustic impedance will change upon the contact of the sensor to the tested soft tissue, resulting in a shift of its resonance frequency. Therefore, the shift of the frequency can be measured as an indicator of the tissue elasticity. However, one problem associated with this frequency shift measurement technique is that the shift is affected by other factors such as the shape of contacted surface and

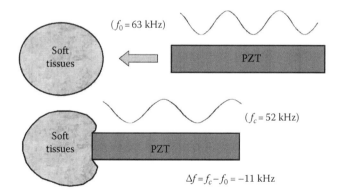

FIGURE 6.1 Illustration of resonant frequency shift measurement.

the contact force. If a spherical indenter is used, the combination of frequency shift method and the indentation method can lead to a new measurement method where only the contact force and the frequency shift are needed for extracting Young's modulus from the tested tissues (Jalkanen 2010):

$$\frac{\partial F}{\partial \Delta f} \propto \frac{E}{\rho} \tag{6.1}$$

where
  F is the contact force
  Δf is the frequency shift of the sensor
  E is Young's modulus
  ρ is the tissue density

This method has been broadly applied to the measurement of elasticity of various soft tissues, including skin (Takei et al. 2004), eye (Eklund et al. 2003), breast (Murayama et al. 2008), prostate (Eklund et al. 1999), liver (Kusaka et al. 2000), bladder (Watanabe et al. 1997), lymph node (Miyaji et al. 1997) and ovum (Murayama et al. 2004), and thus its potential in medical applications has been demonstrated (Lindahl et al. 2009). However, contact is necessary for measurement using a tactile sensor, and it can only measure the elasticity of tissues near the probe. The probing depth is limited, so it is unsuitable for testing inner body tissues or deep regions of a tissue. In spite of these limitations, recently a commercial device called Myoton (Myoton AS, Tallinn, Estonia) has become available using the principle of resonant frequency shift. Because of its portability, Myoton has recently been used for the assessment of various tissues, with its most updated application for muscle assessment in microgravity environments (Schneider et al. 2015).

6.3 VIBRO-ACOUSTIC SPECTROGRAPHY AND HARMONIC MOTION IMAGING

According to the theory of nonlinear acoustics, when two ultrasound beams with different frequencies interact with each other and assume their intensity is large enough for generating nonlinear effects, such interaction will generate new waves with their difference frequency (beat frequency) and sum frequency (Zheng et al. 1999a). The intensity of the new wave depends on a number of factors, including the elasticity of the material, and it has been used for characterising different materials for many years.

Fatemi and Greenleaf (1998, 1999a,b) first proposed a technique named vibro-acoustic spectrography or vibro-acoustography to use this nonlinear acoustic effect for biological tissue assessment (Fatemi and Greenleaf 1998, 1999a,b). They used two confocal ultrasound transducers to transmit two beams with slightly different frequencies to interact with each other at the focal zone, and the difference frequency was generated and detected using a hydrophone placed on the tissue surface (Figure 6.2). They demonstrated that the technique could be used to measure tissue elasticity (Fatemi and Greenleaf 1999a,b) and study breast cancer (Fatemi et al. 2002), prostate cancer (Mitri et al. 2009) and many others (Urban et al. 2011). Instead of using a single difference frequency in its initial version, a method using variable difference frequencies has also been investigated to detect the resonance frequency of the tissue (Urban et al. 2006). While it is a promising technique to detect tissue elasticity, one limitation of this technique is that it requires a hydrophone to be placed on the tissue surface to detect the

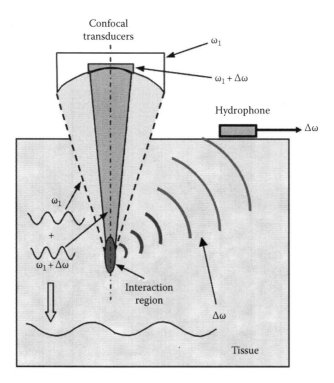

FIGURE 6.2   Illustration of vibro-acoustic spectrography using the interaction of two ultrasound beams with slightly different frequencies and a hydrophone to detect the wave with the beat frequency.

wave with the difference frequency. It would be ideal if a single ultrasound probe is used with all the transducers integrated.

Konofagou and Hynynen (2003) proposed a new technique along this direction, named 'harmonic motion imaging' (Konofagou and Hynynen 2003). They first used two confocal ultrasound transducers to generate two ultrasound beams to interact with each other to generate a low-frequency vibration, i.e. harmonic motion, with the difference frequency as in vibro-acoustography. Instead of using a hydrophone, they used an ultrasound transducer (7.5 MHz) to detect the low frequency tissue vibration to get tissue elasticity, which is the main difference between these two techniques (Figure 6.3). Later, an investigation was conducted to use a single linear ultrasound phased array to induce harmonic motion in the tissue (Heikkila and Hynynen 2006). They also demonstrated that using an amplitude-modulated ultrasound beam can also induce harmonic motion in tissue, and this further simplified the measurement system (Maleke et al. 2006; Maleke and

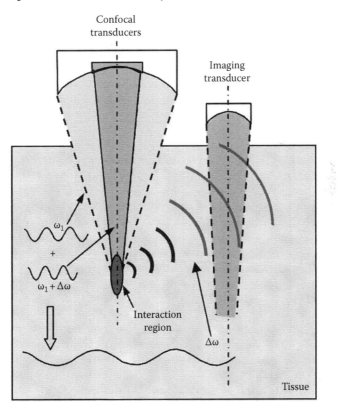

FIGURE 6.3 Illustration of harmonic motion imaging where the wave with the beat frequency is detected using an ultrasound imaging beam.

Konofagou 2008). The feasibility of this technique for various applications, including measurement of the viscoelastic parameters of tissues, has been demonstrated (Vappou et al. 2009).

6.4 DYNAMIC HOLOGRAPHY

Another category of indirect measurement of tissue elasticity uses one probing wave, which can approach the detection point noninvasively. The probing wave can detect certain physical characteristics of the tissue that are related to its elasticity. Therefore, it is possible to detect the elasticity of the tissue using the physical characteristics that the probing wave obtains. Frequently used waves for noninvasive detection mainly include electromagnetic and mechanical waves. The propagation of electromagnetic wave in materials is mainly related to factors including the dielectric permittivity and the magnetic permeability, which are not directly related to the mechanical properties of the material. Therefore, they are not appropriate for the detection of soft tissue elasticity. However, the propagation of mechanical waves is very closely related to the mechanical properties of materials, and therefore they can be used as a suitable tool for probing the soft tissue elasticity. Avehaus et al. used a dynamic holographic endoscopy technique to observe the response of gastric tissue after touching it with a guide wire (Avikainen et al. 1999; Avenhaus et al. 2001; von Bally et al. 2002; Pedrini et al. 2003). They found that the patterns of interferograms were different when elasticity of the tested tissue was different, so that this method could be used to detect the change

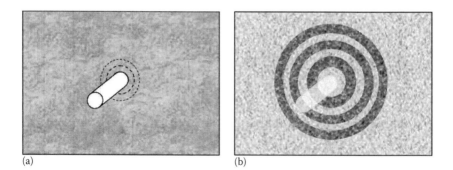

(a)   (b)

FIGURE 6.4 Illustration of holographic image of gastric tissue. (a) Optical image of tissue surface disturbed by a probe, such as a guide wire used in endoscopic imaging. (b) Corresponding speckle correlation patterns obtained, showing wave propagation patterns that may change according to the tissue elasticity and its distribution.

in tissue elasticity (Figure 6.4). Although this method can be used for the detection of difference in elasticity, it is a qualitative method, which is difficult for objective diagnosis. Further investigation is required in this area to extract quantitative elasticity-related parameters to show its potential for the diagnosis of gastric cancers using endoscopy.

While dynamic holography shown in Figure 6.4 mainly senses the elasticity of superficial tissues, shear wave propagation can happen inside tissue, thus providing information on the elasticity of tissues at different depths. In this method, the shear wave can be generated by various methods including mechanical stimulation and acoustic radiation force, and when it propagates into the tissue, medical imaging methods such as ultrasound, MRI or optical imaging can be used to track it. The propagation speed can be measured and used as the parameter to estimate the tissue elasticity. This area has developed rapidly during the last two decades, and thus more details on the principle and on technical state-of-the-art developments are introduced in the following sections.

6.5 SHEAR WAVE PROPAGATION METHOD

6.5.1 Basic Principle of Shear Wave Method

In a homogeneous, isotropic, pure elastic solid material, mainly two types of mechanical waves can propagate, the first one being a compressional wave and the other a shear wave (Raghavan and Yagle 1994). The speed of these two waves in solid is (Parker et al. 2011)

$$C_L = \sqrt{\frac{K + \frac{4}{3}\mu}{\rho}} \qquad (6.2)$$

$$C_s = \sqrt{\frac{\mu}{\rho}} \qquad (6.3)$$

where
  C_L and C_s are the speeds of the compressional wave and shear wave, respectively
  K is the bulk modulus
  μ is the shear modulus
  ρ is the density of the material

K and µ are related through Poisson's ratio:

$$K = \frac{2\mu(1+v)}{3(1-2v)} \qquad (6.4)$$

where v is Poisson's ratio of the material. In elastic material, another important mechanical parameter is Young's modulus E, which has a relationship with the shear modulus given by

$$E = 2\mu(1+v) \qquad (6.5)$$

If we assume that the material is incompressible, i.e. having Poisson's ratio of approaching 0.5, the relationship becomes

$$E \approx 3\mu \qquad (6.6)$$

This relationship has been frequently adopted in the measurement of soft tissue elasticity using the shear wave method. In the following description, we also assume this relationship unless explained otherwise. In a soft tissue, because of its high incompressibility (when deformed such as wave propagation, the fluid inside the tissue does not have the time to move in or out), K is much larger than µ, and therefore the speed of a compressional wave is much larger than that of a shear wave. For example, the mechanical wave used in traditional ultrasound imaging is a type of compressional wave with a propagation speed of ~1540 m/s. However, for the shear wave that is used in tissue elasticity measurement, the propagation speed ranges from several to several tens of meters per second, which is much smaller than that of ultrasound.

In general, the compressional wave propagates in the form of a longitudinal wave, for which the direction of particle vibration is the same as the propagation direction of the wave (Figure 6.5a). For a shear wave, it is normally a transverse wave, for which the vibration direction of particles is perpendicular to the propagation direction (Figure 6.5b). However, this is not absolutely true, as in exceptional cases a shear wave can also propagate in the form of a longitudinal wave (Raghavan and Yagle 1994). For example, when a solid vibrator vibrates on the surface of the tested tissue in the vertical direction, the shear wave propagating

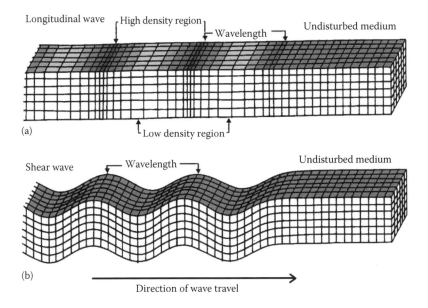

FIGURE 6.5 Illustration of longitudinal (a) and shear (b) waves in a medium with the shear wave only disturbing the meshes while the longitudinal wave also compressing the fluid phase in medium, if any.

in its near field in the vertical direction underneath the vibrator is in the form of a longitudinal wave (Raghavan and Yagle 1994). As the difference of the bulk modulus is small for most soft tissues, which is normally within one order of magnitude (10^9–10^{10} Pa) (Sarvazyan et al. 1998), the difference of compressional wave propagation speed (mostly, ultrasound) in various soft tissues is not very obvious. The situation is different for a shear wave, as its speed is mainly decided by the shear modulus of the tissue, which is not much related to the fluid phase in the tissue. The difference of shear modulus among tissues is quite large, which can be of several orders (10^3–10^8 Pa) (Sarvazyan et al. 1998). Therefore, the difference of shear wave speed in soft tissues can be quite large. For the same type of tissues, the bulk modulus can be almost the same between different pathologies, but the shear modulus can vary in a much larger range. For example, the bulk modulus of different breast tissues is ~2000 MPa; however, in the case of Young's modulus, it is about 20 kPa for fat tissues but 100 kPa for fibrotic tissues and carcinomas, so the difference is quite large (Krouskop et al. 1998; Wells and Liang 2011). Therefore, the measurement of shear wave speed can be used as a method to detect the pathology of the tissue.

After the shear wave speed is measured, Young's modulus of the tested tissue can be estimated as

$$E = 3\rho C_s^2 \qquad (6.7)$$

In this formula, a tissue density of 1000 kg/m³ can be used for the calculation. Then, the next question for the elasticity measurement using the shear wave method is how to generate the shear wave and how to detect it in the soft tissues, which is described in the following section.

6.5.2 Measurement Techniques of Shear Wave Propagation

According to types or generation methods of the shear wave used in elasticity measurement, the measurement techniques are mainly divided into the following three categories.

6.5.2.1 Sonoelastography

Sonoelastography utilises the propagation of a continuous wave for the measurement of tissue elasticity. Lerner et al. were inspired by hand palpation and first proposed a vibration amplitude–based elasticity imaging method (Lerner and Parker 1987; Lerner et al. 1988, 1990; Parker et al. 1990). In their method, they used a mechanical vibrator placed on the tissue surface (Figure 6.6a). The low surface vibration (20–1000 Hz) can then propagate into the internal parts of the tissue and induce the vibration of particles. With the combination of Doppler measurement using ultrasound, the vibration speed in the tissue can then be measured, which can be used

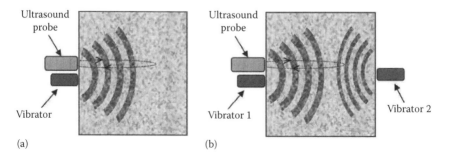

FIGURE 6.6 Illustration of sonoelastography (a) using a single vibrator and (b) using two vibrators with different frequencies. Tissue vibration is detected using ultrasound Doppler imaging.

as a factor to calculate the vibration amplitude. If tissue is assumed to be homogeneous, the distribution of vibration amplitude in the tissue should have some eigenmodes. While the homogeneity of elasticity is broken by the existence of a local pathology, the distribution of vibration amplitude will be significantly changed, so it is possible to detect such a change qualitatively. This method was initially proposed for the imaging of tissue elasticity, but it is difficult to be used as a quantitative method to calculate the value of the elastic modulus. Therefore, the details are not given here.

By simplifying the model, Krouskop et al. proposed a method to quantitatively measure the elastic modulus of muscular tissue by measuring the gradient of vibration amplitude inside the tissue (Krouskop et al. 1987). Later, Yomakoshi et al. proposed to measure the amplitude and phase of the shear wave using Doppler ultrasound (Yamakoshi et al. 1990). Based on the phase distribution, the wavelength can be measured, and then the speed can be calculated with the frequency information. The speed can then be used to calculate the tissue elasticity. In detail, for the ultrasound Doppler signal backscattered from the tissue, a cross-correlation-based demodulation is performed and two signals are obtained from the quadratic detection (Yamakoshi et al. 1990):

$$d_1(t) = K \left\{ \cos\phi \left[J_0(m_f) + 2\sum_{n=1}^{\infty} J_{2n}(m_f)\cos 2n(\omega_b t + \phi_b) \right] \right.$$

$$\left. - 2\sin\phi \sum_{n=0}^{\infty} J_{2n+1}(m_f)\sin(2n+1)(\omega_b t + \phi_b) \right\}$$

$$d_1(t) = K \left\{ \sin\phi \left[J_0(m_f) + 2\sum_{n=1}^{\infty} J_{2n}(m_f)\cos 2n(\omega_b t + \phi_b) \right] \right.$$

$$\left. - 2\cos\phi \sum_{n=0}^{\infty} J_{2n+1}(m_f)\sin(2n+1)(\omega_b t + \phi_b) \right\} \tag{6.8}$$

where

ϕ is the phase induced by propagation

$J_i(x)$ is the ith-order Bessel function

ω_b is the low-frequency vibration with phase of ϕ_b

K is a system-related constant

m_f is a Doppler modulation factor, the value of which is related to the vibration amplitude as

$$m_f = \frac{2\omega_0 \xi_0}{C_L} \tag{6.9}$$

where
ω_0 is the central frequency of the ultrasound
C_L is the ultrasound speed
ξ_0 is the vibration amplitude

From Equation 6.8, we can see that the demodulated signal has DC and AC parts with the coefficients decided by the modulation factor m_f and the Bessel functions. Based on the relationship between the coefficients at different harmonic frequencies and values of Bessel functions, the modulation factor m_f can be obtained for calculating the vibration amplitudes at different sites inside the tissue. Another important factor is the phase ϕ_b of vibration at that point, which can be obtained from the phase of the signal at the fundamental frequency. When phase information is obtained and monitored continuously in time, the wave propagation can be tracked and then the shear wave speed can be calculated for obtaining the shear modulus of the tissue.

In addition to the method of stimulation generated from a single source, the shear wave speed can also be obtained from the speed of a crawling wave generated from two vibration sources with a beat frequency (Wu et al. 2004b, 2006). Two vibrators with similar but slightly different frequencies are placed at the two sides of the tissue, and the interaction of the two shear waves produces a crawling wave, the speed of which is smaller than that of the original one (Figure 6.6b). Then the crawling wave will propagate in the direction from the source of higher frequency to that of lower frequency. The propagation speed for this crawling wave is proportional to the frequency of the original shear wave (Wu et al. 2004):

$$V_{crawl} = \frac{\Delta\omega}{2\omega_0} C_s \tag{6.10}$$

where
ω_0 and $\Delta\omega$ are the fundamental frequency and the frequency difference
C_s is the speed of the original shear wave

As the beat frequency is much smaller than the fundamental frequency, the crawling wave speed is also much smaller than that of the shear wave. Therefore, the propagation of the crawling wave can be easily tracked using conventional imaging methods such as ultrasound imaging (Hoyt et al. 2007, 2008a).

A conventional ultrasound imaging device with a module for Doppler flow measurement can be modified slightly for sonoelastography when additional vibration sources are adopted for generating the shear wave. The shortcoming of the detection using the Doppler effect is that the results can be easily affected by the quality of the Doppler signal. In practical measurements, refraction and reflection of the shear wave can make the real form of mechanical wave in the tissue quite complicated because of complex boundary conditions and the limited vibrator size (Catheline et al. 1999a). How to place these vibration probes and the ultrasound transducer beside the tissue (especially when two probes are needed) is also a critical question that needs to be carefully considered; it may need a special design in practical measurement. Because of these difficulties, most sonoelastographic measurements are still in the experimental stage, and this technique has not been commercialised yet.

6.5.2.2 Transient Elastography

To explain the problem of elasticity measurement using a continuous shear wave, Catheline et al. explored the propagation of a low-frequency (10–300 Hz) shear wave in biological phantoms and real soft tissues in detail (Catheline et al. 1999a,b). They placed the vibration source at one side of the phantom and the ultrasound transducer at the other side just facing the vibrator (Figure 6.7a). M-mode ultrasound was used to track the ultrasound propagation. During the experiment, when single-frequency continuous ultrasound was used, there was a large error induced by refraction, reflection and effects of the compressional wave. However, if pulsed ultrasound was used, the effect induced by these situations could be significantly reduced, thereby greatly improving the accuracy of the measurement. The method that uses pulsed ultrasound for the measurement of shear wave speed and thus the elasticity of the soft tissue is called transient elastography (TE).

As explored by Catheline et al. (1999a), at the beginning, the ultrasound transducer was placed on the other side of the vibrator for detecting the shear wave. As it was inappropriate to apply this method in clinical situations, Sandrin et al. proposed the reflection ultrasound–based transient elastography (Sandrin et al. 2002) (Figure 6.7b). In this method, the ultrasound transducer itself is used also as the vibrator for producing the shear

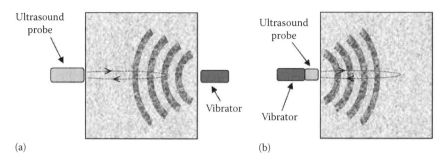

(a)   (b)

FIGURE 6.7   Illustration of transient elastography, where ultrafast ultrasound imaging is used to track the propagation of shear waves in tissue. (a) Vibration source and ultrasound probe are arranged at the opposite sides of tissue. (b) Vibration source and ultrasound probe are combined and arranged on the same side of tissue.

wave, and then M-mode ultrasound is used to track the propagation of the shear wave (Figure 6.8). The difference between these two methods is that in the transmission mode the ultrasound transducer does not move with the vibrator so that the displacement it measures is just the displacement of the tissue; however, in the reflection mode, the displacement measured is affected by the movement of the transducer itself, so it needs some compensation. A fixed point in the tissue such as the soft tissue–bone interface can be used as the reference point in the displacement compensation. Another scheme is to select a deep enough point in the tissue so that one can assume that the shear wave vanishes here after propagation through a long distance due to attenuation. Then the measured displacement can

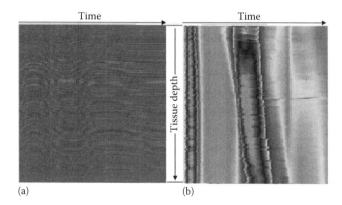

(a)   (b)

FIGURE 6.8   Display of signals detected by ultrafast ultrasound imaging. (a) M-mode display of radio frequency ultrasound signal. (b) Trace of wave propagation obtained by processing signals shown in (a).

be treated as the movement of the transducer itself and can be used for the compensation. If these conditions cannot be well met in the study, a further step of differentiation, after compensation for calculation of strain, can reduce the error of measurement (Sandrin et al. 2002). It has been demonstrated that the reflection-mode transient elastography technique could be successfully applied to the measurement of soft tissue elasticity *in vivo* to differentiate the significant change of tissue elasticity in the biceps between relaxed and contraction states (Sandrin et al. 2002).

In addition to placing the ultrasound transducer under the vibrator or just using the ultrasound transducer as both the vibrator and signal receptor, another mode of operation is to place the ultrasound transducer along the propagation direction at the tissue surface. Two observation A-lines are used to detect the time for the propagation of shear wave from one place to another. The time of flight for the shear wave to travel a fixed distance between the two lines is used to calculate the wave speed and then the elasticity of soft tissues. Wang et al. utilised this technique to measure the elasticity of blood vessels and muscles (Figure 6.9) (Wang and Zheng 2009; Wang et al. 2010a). Through a maximum isometric voluntary contraction

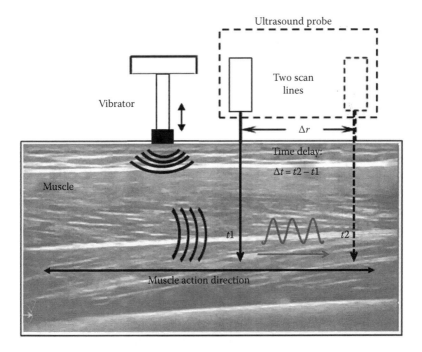

FIGURE 6.9   Illustration of shear wave propagation method with two ultrasound beams perpendicular to the wave propagation direction.

test, they found, surprisingly, that the change of the muscle stiffness between relaxation and contraction states could be as large as hundreds of times (Wang 2011; Wang and Zheng 2011; Wang et al. 2014).

A commercial measurement system, the Fibroscan® (by Echosens, Paris, France, operated as a subsidiary of Inner Mongolia Furui Medical Science Co. Ltd., China), has been developed for the measurement of liver tissue elasticity, based on the transient elastography technique. The measured liver stiffness can serve as an important parameter to indicate the fibrotic information of the liver, which is very useful for the screening and diagnosis of liver fibrosis (Sandrin et al. 2003; Castera et al. 2005; Foucher et al. 2006). Fibroscan has become a very useful clinical tool for the noninvasive assessment of liver status. However, Fibroscan is an ultrasound system different from traditional ultrasound imaging machines, and it does not have imaging guidance for the operation during the assessment. As is known, the liver is located in the thoracic cage where the ribs may prevent the penetration of ultrasound and shear waves deep into the liver. Only the part between two consequent ribs can be used for the detection, and, furthermore, the anatomy of liver is quite complicated with blood vessels and bile ducts, which may render the results of the measurement meaningless. Therefore, experience is necessary for the operator of Fibroscan to correctly avoid big vessels or ducts and locate the region of interest for measurement, which can be better conducted with some type of guidance, to improve the test reliability and accuracy. Zheng et al. proposed an ultrasound image guiding method to help locate the measurement region before transient elastography measurement (Figure 6.10) (Zheng et al.

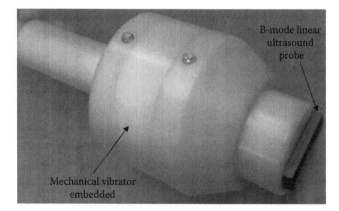

FIGURE 6.10 B-mode ultrasound probe integrated with a vibrator for imaging and elasticity measurement of liver.

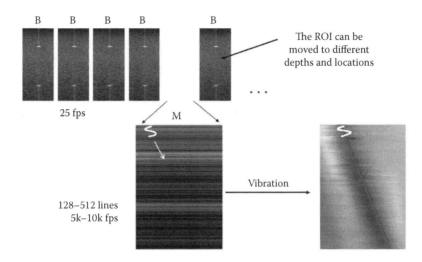

FIGURE 6.11   Illustration of the working principle of tissue elasticity measurement with real-time ultrasound image guidance.

2010; Mak 2013; Mak et al. 2013). Ultrasound imaging of the liver is first performed, and a homogeneous region without large vessels and ducts can be selected for placing the A-line in order to define the region of interest. After that, M-mode ultrasound signals are collected to measure the speed of the shear wave (Figure 6.11). The transient B-mode elastography technique guided by ultrasound imaging can reduce the measurement variability and operator dependence for the liver stiffness measurement. More clinical trials on this system are still going on.

6.5.2.3 Acoustic Radiation Force-Based Elasticity Imaging Techniques

In addition to the method of producing shear waves using a mechanical vibrator placed on the tested tissue surface, acoustic radiation force also can be used as an alternative technique to induce the shear wave. When an acoustic wave is reflected or absorbed during its propagation, force will be generated, which is called the acoustic radiation force (Torr 1984). In soft tissues, the acoustic radiation force is normally produced at the point of focus, where the absorption of acoustic energy is maximum. Acoustic radiation force has the same direction as the propagation of the acoustic wave, which makes the particle of the tissue vibrate in this direction. Then the produced shear wave propagates in a direction perpendicular to the direction of vibration. The phenomenon of acoustic radiation force was discovered in the early 1900s, and it was scientifically defined in Rayleigh's seminal paper 'The pressure of vibrations' (Rayleigh 1902; Sarvazyan et al. 2010).

It was not until 1998 that Sarvazyan et al. (1998) proposed it as a way of producing shear waves for the measurement of tissue elasticity.

Sarvazyan et al. (1998) made a thorough analysis of the vibration and shear wave induced by acoustic radiation force and proposed three potential ways to utilise this phenomenon for soft tissue elasticity measurement. The first is to measure the time to reach maximum displacement in the focal point. This time is related to the shear wave velocity, so a direct measure of the time can be used to probe the shear wave velocity in tissues. This method is not very straightforward. The second method is to use the value of displacement itself in the focal point because it is inversely proportional to the square root of shear modulus. The third method is to observe the wave front of the propagating shear wave and then calculate its velocity for calculating the shear modulus (Figure 6.12). As it is a single-point measurement and the measurement can be affected by a number of factors for the first and second measurement methods, only the third method has been explored in detail and broadly applied in clinical situations for the measurement of soft tissue elasticity. In the direction of propagation of shear wave, two points can be set up to trace the passage of the wave in order to measure the propagation speed (Figure 6.12b).

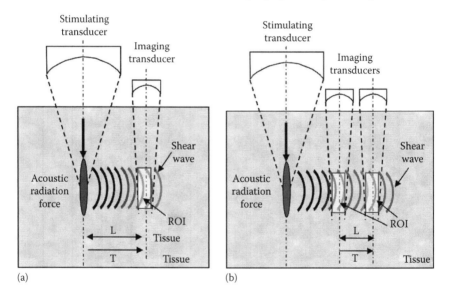

FIGURE 6.12 Illustration of acoustic radiation force-based shear wave propagation method, with the speed calculated as L/T, with L representing the distance and T the time. (a) Using a single beam for detection. (b) Using two beams at different locations for detection.

One advantage of this method is that the location where acoustic radiation force is generated need not be accurately known. Chen et al. proposed a frequency-dependent model of wave speed to measure the essential elasticity and viscosity of the tested object (Chen et al. 2004):

$$C_s(\omega) = \sqrt{\frac{2(\mu^2 + \omega^2\eta^2)}{\rho(\mu + \sqrt{\mu^2 + \omega^2\eta^2})}}$$

(6.11)

where

$\omega = 2\pi f$ is the angular frequency

μ is the shear modulus (in Pa or kPa)

η is the viscosity coefficient (in Pa·s or kPa·s)

Through measurements at multiple frequencies and the relationship between shear wave speed and frequency, as shown by Equation 6.11, two parameters, i.e. shear modulus and the viscosity coefficient, can be extracted by regression, which is called as 'shear wave dispersion ultrasound vibrometry' (Chen et al. 2009). Later, the sources of errors for using this method were discussed (Urban et al. 2009), and the accuracy of measurement was verified by standard indentation test on gelatin phantoms as test objects (Amador et al. 2011). A preliminary test was conducted on the prostate tissue *in vitro* to demonstrate the method's feasibility in measuring tissue viscoelasticity for the early detection of prostate cancer (Mitri et al. 2011). The advantage of this method is that two distinct parameters can be used to represent the intrinsic elasticity and viscosity of the tissue separately, which can then describe the mechanical properties of the tissue in a better way.

In order to measure the propagation of a shear wave in space, a fast imaging method is needed. Traditional ultrasound imaging (normally <100 frames/s) is not fast enough to track the propagation of a shear wave, although a multiple measurement scheme can be designed to virtually increase the imaging speed (Nightingale et al. 2003). The method first used a single A-line selected at a certain place of the field to detect the passage of the shear wave based on high-speed M-mode operation. A long enough observation time was assumed, so it was assumed that the shear wave was successfully detected at the selected place using M-mode imaging. The stimulation was then regenerated and the observation was changed to another place so that multiple space measurements were possible. Strict synchronisation and integration of these data could lead to a virtual observation of the shear wave propagation in the field, and the space versus time distribution of the vibration displacement could be used to measure the shear wave propagation speed (Nightingale et al. 2003). This technique could be easily implemented in conventional ultrasound machines with minor modifications. As for ultrasound systems in the commercial market, the Acouson S2000 system from Siemens includes such a module, which can measure the shear wave speed induced by acoustic radiation force in a region of interest selected in the tested soft tissue, called 'virtual touch quantification' if it is a single point measurement and 'virtual touch IQ' if it is used for imaging. From the ultrasound B-mode image, a region of interest can be selected. Then, vibration induced by acoustic radiation force is produced at its nearby field, and the time for the shear wave to propagate from the generation point to the region of interest will be used to calculate the wave speed, which is an average value over the region of propagation (Figure 6.13).

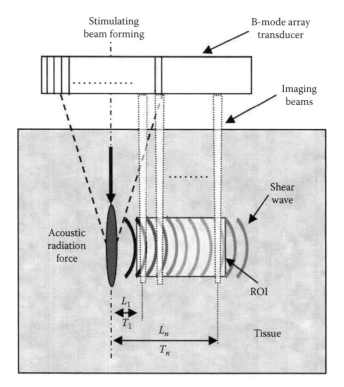

FIGURE 6.13   Illustration of generating acoustic radiation force using an array probe, which is also used as the imaging probe to detect the shear wave speed at multiple points within the region of interest.

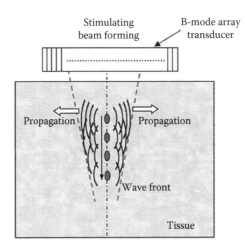

FIGURE 6.14   Illustration of supersonic shear imaging using acoustic radiation force. After the shear wave is generated, ultrafast ultrasound imaging is then used to monitor its propagation.

In addition to the single-source measurement, a series of vibration sources along the depth direction can also be produced by acoustic radiation force for monitoring the shear wave propagation (Figure 6.14). For example, supersonic shear imaging (SSI) was first proposed by a French group to produce a line of sources to form a shear wave with almost a plane wave front (Bercoff et al. 2004a). A company (Supersonic Imagine, Aix-en-Provence, France) was established for the commercialisation and marketing of this technology. The principle of the SSI technique is to rapidly produce a line of vibration sources and then to detect the spatial time displacement distribution caused by the source vibration using a specialised ultrafast (>5000 frames/s) ultrasound imaging system (Sandrin et al. 1999). Various algorithms such as differential or time-of-flight techniques have been proposed for extracting the spatially distributed shear wave speed (Bercoff et al. 2004a; McLaughlin and Renzi 2006a,b). This method is called 'supersonic' because the effective 'speed' of the vibration source formation along the source line is faster than the propagation of the shear wave. This type of stimulation can ensure that the propagation of shear wave happens after the generation of the vibration sources. The ratio between the 'motion' of generated vibration source and the shear wave speed is called the 'Mach number', which is an important factor for this technique. The Mach number can be varied to adjust the angle produced by the two propagation wave fronts at both sides of the source line.

The angle of the propagation fronts can be varied obviously when the Mach number is in the range 1–5; however, the wave fronts are nearly parallel when the Mach number is larger than 5.

A shear compounding technique can be designed for SSI in order to improve the image quality (Bercoff et al. 2004a). Shear compounding is achieved by insonifying the tissue with different angles using different Mach numbers. Another issue is that for a single SSI imaging, although the shear wave speed can be measured at regions away from the source, the measurement near the source region is not accurate and this needs to be specially treated (Bercoff et al. 2004a). One simple solution is to conduct the imaging twice or more in order that the source regions at the beginning of the test are placed at the imaging regions of later tests. After integration, then a full range of elasticity distribution in the whole field can be successfully obtained using SSI (Tanter et al. 2008). SSI can be used to measure the elasticity of multiple points or imaging of tissue elasticity, and the results are less vulnerable to other factors, so it has the potential to be broadly applied to clinical situations. Preliminary clinical studies have been reported for a number of tissues including breast (Tanter et al. 2008; Athanasiou et al. 2010), liver (Muller et al. 2009), muscle (Gennisson et al. 2010; Nordez and Hug 2010; Shinohara et al. 2010; Arda et al. 2011), thyroid nodule (Sebag et al. 2010), brain (Mace et al. 2011) and cornea (Tanter et al. 2009).

One important issue related to the use of acoustic radiation force is its safety. Sarvazyan et al. (1998) roughly pointed out that the spectral density used would be similar to that for conventional ultrasound imaging; however, it would last for a longer time when used for a source of vibration. But still it did not reach the limit of producing significant biological effect, and therefore is safe (Sarvazyan et al. 1998). Palmeri et al. studied the thermal effect of acoustic radiation force on soft tissues using computer simulation and experiments (Palmeri et al. 2004; Palmeri and Nightingale 2004; Fahey et al. 2005). They concluded that the temperature effect induced by acoustic radiation force is significantly dependent on the attenuation coefficient of soft tissue and the excitation frequency, and the acoustic radiation force could be generally used as a source for imaging under the safety level. However, care should be exercised to avoid too frequent or too long exposure, and a balance should be made between imaging speed, size of the detection field and interval of the scan lines. For single-value stiffness measurement, the excitation is done only once by the focused ultrasound, so it is generally safe. For SSI, it was computationally safe (Bercoff et al. 2004a). Athanasiou et al. reported the use of a clinical ultrasound system using SSI for the imaging of breast. The mechanical index (MI) was 1.4, I_{SPTA} was 603 mW/cm² and the thermal index (TI) was 0.48 at 1 Hz SSI frame rate, all of which were smaller than those safety values regulated by the FDA (MI limit: 1.9; I_{SPTA} limit: 720 mW/cm²; TI: 6) (Athanasiou et al. 2010). In summary, under single-time excitation or multiple excitations with a not-too-high frequency, imaging using acoustic radiation force is safe, but if the excitation is produced with a high continuous speed, safety issues must be considered.

6.6 ADVANTAGES, PROBLEMS AND FUTURE PERSPECTIVES

Tissue elasticity measurement using shear wave as the probe has a very important advantage in that it can locally measure the mechanical properties of soft tissue at different depths. Furthermore, the measurement is less affected by geometry, size and the surrounding tissues at the test point. It is also less affected by the probe itself, so the operator just needs to place it on the test site and let the shear wave successfully propagate through it, and the system will automatically measure and obtain the results. Thus, effects from the environment and due to manual operations are significantly reduced, so the test results become more reliable and accurate. There is high potential for soft tissue elasticity measurement and imaging *in vivo* as a clinical method.

In this chapter, the ultrasound method has been introduced as a main technique for the detection of shear wave propagation in tissues. The advantages of using ultrasound for detection include the easy accessibility of ultrasound devices and high speed for real-time imaging. However, limitations are also obvious when ultrasound is used for those parts covered by bones with strong reflection, such as the brain. One solution is to use other medical imaging methods such as MRI. As known in this field, MR elastography (MRE) has become quite an important technique for the measurement of soft tissue elasticity (Muthupillai et al. 1995; Ringleb et al. 2007; Di Ieva et al. 2010; Mariappan t al. 2010). Its applications in brain tissue imaging are unique and have attracted much attention from both academics and doctors in making this technique a real clinical diagnostic option. Readers can refer to some excellent research and review papers in this field for further information (Green et al. 2008; Kruse et al. 2008; Sack et al. 2008; Di Ieva et al. 2010). An emerging direction of MRE is to use acoustic radiation force to induce vibration inside tissue (McDannold and Maier 2008; Souchon et al. 2008), so that tissue vibration can be better controlled.

For the ultrasound-based shear wave propagation method, there are still some problems or issues that need to be further solved or addressed. First, most of the measurements were based on a pure elastic model and neglected the viscous properties of the tissue. In fact, soft tissues have quite obvious viscous properties, and one simple phenomenon is the frequency-dependent shear wave speed, which cannot be explained by a simple elastic model. There have been some studies that proposed some specific models to study the viscoelastic properties of the measured soft tissue (Bercoff et al. 2004b; Catheline et al. 2004; Deffieux et al. 2009), also including the shear wave dispersion ultrasound vibrometry (SDUV) method (Chen et al. 2004). However, one remaining issue is how to define simple but effective viscous parameters to be measured and used for clinical diagnosis.

The second issue is the relationship between the change of elasticity and pathology of the tissue structure. For example, in the liver, even without significant change of fibrosis, the stiffness can increase dramatically to be close to or larger than the critical value for diagnosis of fibrosis due to acute hepatitis alone (Cobbold and Taylor-Robinson 2008). There is no consensus on how the acute change in pathology can lead to the change of soft tissue stiffness. It is not enough to just get the stiffness value for diagnosis without knowing the mechanism of changes behind. In future studies, it may be necessary to use biomechanical and physiological models to investigate the relationship between tissue elasticity, tissue pathologies and interfering factors, in order to improve the reliability and accuracy of clinical diagnosis.

Third, most of the current research focuses on the measurement of large tissues, but study of elastic properties also has the potential for studying small tissues such as the skin, digestive tract wall, articular cartilage and eye. Currently, the resolution for the study of elasticity of large tissues (normally >1 mm) is not high enough, and the techniques need to be improved in order to design proper tools for studying these small tissues. This is a very promising field, where new technologies and techniques can be developed for better study of those small tissues. High-resolution imaging methods, such as very high–frequency ultrasound imaging and delicate optical imaging methods can be adopted in mechanical tests to solve the measurement issues in small tissues. One emerging area is to use optical coherence tomography in ultra-high frame rate (20,000 frames/s) to monitor shear waves induced by different sources, including pulsed excitation laser beam (Li et al. 2012a) and air-jet pulse (Wang et al. 2013). Both techniques have demonstrated the feasibility for the measurement of cornea elasticity. Another emerging area uses acoustic radiation force to induce tissue vibration and optical imaging to monitor its propagation (Bouchard et al. 2009b). Because of the high resolution provided by optical coherence tomography, high spatial resolution of elasticity mapping can be achieved, and the measurement can be conducted without any contact.

In summary, the development of various reliable and accurate shear wave measurement techniques for the mechanical characterisation of different types of tissues will undoubtedly bring new opportunities to solve the remaining clinical and biological problems, pointing to a bright future.

CHAPTER 7

Elasticity Imaging and Elasticity Measurement

7.1 INTRODUCTION

As discussed in previous chapters, elastic properties are important material properties that deserve quantitative measurement for tissue characterisation or diagnosis. Depending on the number of measurement points or measurement resolution, two different concepts are used: imaging and measurement. When the purpose is to map the elasticity distribution among the tissues, or to detect a focal disease, it is more appropriate to perform elasticity imaging, or elastography; otherwise, if it is for a single and integral measurement or for detecting a diffused disease, elasticity measurement can be used.

Let us have a close look at the comparison between elasticity imaging and measurement in terms of several different characteristics including requirements for resolution, computational complexity and parameters. The resolution required for imaging is higher than that for measurement, because the former deals with mapping of the elasticity at each point. The purpose of elasticity mapping is to find any abnormal lesion embedded in a possibly homogeneous background, so the resolution of imaging should be at least able to discriminate the lesion from its surroundings (Figure 7.1). Regarding the computational complexity, that of imaging is generally higher than of measurement because multiple point measurement is involved in elasticity imaging. Concerning the parameters used, intrinsic material properties are always used in the measurement because their values may be compared

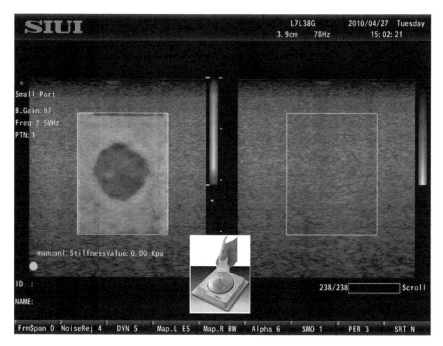

FIGURE 7.1   Ultrasound elasticity imaging of a breast phantom using quasi-static compression.

with those of experiments conducted at different times, on different body parts or in different places. Therefore, the parameters adopted should not vary with the time and place. Only intrinsic material properties meet these specific requirements. However, when the same instruments and the same test protocols are used, these requirements can be relaxed and non-intrinsic parameters can also be used for relative comparison purposes. For elasticity imaging, both intrinsic and phenomenological parameters can be used because the local contrast is the most important factor in the case of detecting focal lesion in elasticity imaging. For example, strain is not an intrinsic parameter, but strain imaging is broadly adopted in ultrasound elastography to detect hard inclusions because it can clearly show the contrast necessary for diagnosis. In the following section, a brief introduction of different elasticity imaging methods and their state-of-the-art development is given.

7.2   ELASTICITY IMAGING TECHNIQUES

7.2.1   Quasi-Static Compression and Strain Imaging

The basic step to use strain as a parameter for elasticity imaging is to first induce a quasi-static (low frequency) compression to the tissue and then measure the local displacements inside it. The gradient of the displacement can then be used as an estimation of the strain for showing the difference of elasticity at different locations of the tissue, assuming a uniform distribution of stress. This method was pioneered by Ophir's group at the beginning of the 1990s (Ophir et al. 1991). Radio frequency signals can be obtained before and after compression, and cross-correlation analysis can be adopted for calculating the displacement induced in the tissue. By assuming a uniform stress level, the strain is inversely proportional to Young's modulus of the tissue so that strain imaging is helpful for detecting a lesion with change of stiffness induced by tissue pathologies.

Quasi-static compression can be performed either externally using a compressor (mostly the ultrasound probe itself) or internally utilising the physiological movement induced by breathing or beating of the heart (deKorte et al. 1997; de Korte and van der Steen 2002; Varghese and Shi 2004; Bae et al. 2007; Swiatkowska-Freund and Preis 2011). For the external stimulation methods, the ultrasound probe can serve as the compressor itself, and freehand compression using the ultrasound probe is the most common method to perform the elastographic imaging *in vivo*. As the most common and easiest method of stimulus in elastography, this mode of elastography was commercialised first in the Hitachi ultrasonic imaging machine called Real-Time Tissue Elastograph, later also available in machines by Siemens and GE, SonixTouch and many other ultrasound machines (Varghese 2009). Alternatively, it is possible to use the physiological stimuli including respiration or cardiovascular pulsation for elasticity imaging. For example, in the intravascular ultrasound elastography, the change of intraluminal blood pressure in a systolic and diastolic cycle is used as an internal source of force to strain the vessel wall. It was found that the near-end diastole is the best period for the strain imaging, because the motion of the catheter is minimal at this moment (de Korte et al. 2002).

A forward problem and an inverse problem can be practised as the solutions of elasticity imaging. The forward problem just measures the displacement, strain or other related parameters and displays them as the contrasting parameter for elastography. However, the method with the inverse problem not only measures strain but also tries to reconstruct the intrinsic mechanical properties such as Young's modulus based on stress–strain relationship and boundary conditions (Kallel and Bertrand 1996). It is obvious that the inverse problem is a better solution for elasticity imaging because it adopts the intrinsic mechanical properties, but it is much more complicated and impractical for clinical applications; nowadays, the forward problem solution has become the mainstream method in quasi-static elasticity imaging.

Three concepts have been introduced in axial strain imaging to discuss the factors affecting the image quality (Varghese and Ophir 1997; Varghese et al. 2001). These are the contrast transfer efficiency (CTE), the strain filter and, finally, the image characterisation. CTE indicates the process of deformation (quasi-static compression) with boundary conditions that converts the contrast of elastic moduli between an inclusion and the background into a contrast of strain, i.e. the contrast of elastic moduli is not perfectly converted to the contrast of strain. CTE is defined as (Varghese et al. 2001)

$$CTE = \frac{C_s}{C_t}$$ (7.1)

where

$C_s = s_1/s_2$ is the strain contrast
$C_t = \mu_1/\mu_2$ is the shear modulus contrast of the two regions in comparison

Simulation results show that CTE is optimal for a stiffer inclusion in a softer background than a softer inclusion in a stiffer background (Kallel et al. 1996). The second concept is termed as a 'strain filter' defined in a statistical way, which basically discusses the difference between ideal strain and its variation caused by limitations of ultrasound data acquisition systems and signal processing parameters. This strain filter can be treated as a bandpass filter in the strain domain, limited by the ultrasound noise effect in the low-strain region and by severe signal decorrelation in the large-strain direction. The third concept is the contrast-to-noise ratio in the final elastogram (CNR$_e$), which combines the first two, CTE and the strain filter. CNR$_e$ is defined as (Varghese et al. 2001)

$$CNR_e = \frac{2(s_1 - s_2)^2}{\sigma_{s_1}^2 + \sigma_{s_2}^2}$$ (7.2)

where

s_1 and s_2 are the mean strain
$\sigma_{s_1}^2$ and $\sigma_{s_2}^2$ are the strain variances in the two regions in comparison, respectively

The strain values in the numerator are closely related to CTE, as shown in Equation 7.1, and the denominator is computed using the strain filter.

Furthermore, it should be noted that there is a nonlinear variation of CNR_e induced by a potentially nonlinear relationship between stress and strain in biological tissues, especially when a large strain exists (Varghese et al. 2001).

In addition to axial strain, there are other parameters that can be used for elasticity imaging. For example, Konofagou et al. (1998, 2001) proposed the concept of Poisson's ratio imaging (Konofagou and Ophir 1998; Konofagou et al. 2001). Although the contrast using the Poisson's ratio as a parameter for imaging may be limited because of the limited range of Poisson's ratio in soft tissues (0–0.5), it has some advantages, such as taking the lateral displacement into account during the compression, which makes the estimation of the axial strain more accurate (Konofagou and Ophir 1998). Furthermore, the longitudinal observation of Poisson's ratio after compression can be used as an indirect way to assess the water-holding capacity of unbound water in normal and pathological tissues such as oedematous tissues (Konofagou and Ophir 1998). Because of the existence of unbound water, there is a hydrodynamic process undergoing in the tissue after the external compression. Therefore, the compressibility of the tissue changes with time, which makes Poisson's ratio change with time.

7.2.2 Dynamic Shear Wave Elastography

In elastography using quasi-static compression with strain imaging, no intrinsic tissue property is obtained from the tested tissues. In applications where both imaging and measurement are preferred, a technique that can achieve both functions is more attractive. The shear wave propagation method, as introduced in Chapter 6, can be used in both elasticity measurement and imaging. The distinct advantage of the wave propagation method is that the measurement is not very sensitive to the boundary conditions. As long as one can measure the propagation speed in a localised region, the shear modulus can be obtained for that region. For a large region, it is easier to measure the average wave speed so that an average Young's modulus can be measured from the examined region. However, if the resolution of wave speed measurement is high enough, this method can also be used as an imaging method, and the contrast between normal and pathological tissues can be naturally adopted as the mechanism for the detection of diseases. Supersonic shear wave imaging (SSI) is such a technique that can be utilised both as an elasticity measurement and imaging tool for the characterisation of soft tissues. For example, measurement of muscle elasticity during isometric contraction can be measured using SSI, which was found to be significantly correlated to the muscle activity level in terms of electromyography (EMG) signal strength (Nordez and Hug 2010). SSI could also be used as an elasticity imaging method to detect the difference of breast tumours (both malignant and benign) compared to its surrounding tissues, which showed a larger contrast for malignant than benign lesions (Athanasiou et al. 2010). It should be noted that if the elasticity measurement is performed based on elastographic images, the operator should know the source of contrast in the elastographic images clearly. For example, if the elastographic image is coming from the quasi-static strain imaging or acoustic radiation force imaging (ARFI, such as virtual touch imaging in the Siemens machine), the results cannot be directly used to get the shear modulus of the tissue for quantitative comparison.

7.3 SUMMARY

Concepts of elasticity imaging and elasticity measurement are not independent; rather, they are closely related. In quasi-static compression elastography, the use of a strain imaging can be used to detect the existence of hard or soft inclusions in a uniform background. With proper measurement of stress and boundary conditions, it is also possible to estimate the distribution of elastic moduli in the tissue, although the whole process is quite complicated and time consuming. With dynamic measurement based on the shear wave propagation, it is easier to achieve both functions of elasticity imaging and measurement. Therefore, this technique is very promising as a versatile clinical modality for a complete characterisation of soft tissue elasticity, for both diagnosis and monitoring of treatment efficiencies. The quasi-static strain imaging method can be developed as a low-cost screening tool for the detection of abnormal lesions such as breast cancer, while the shear wave propagation method can be further developed as a reliable and reproducible tool for quantitative and accurate characterisation and diagnosis of various elasticity change–related pathologies. Readers interested in more details about different tissue elasticity imaging techniques may refer to a number of comprehensive review articles in the field (Gao et al. 1996; Ophir et al. 1996, 1997, 2011; Parker et al. 2005, 2011; Mariappan et al. 2010; Fink and Tanter 2011; Hall et al. 2011; Nightingale 2011; Parker 2011; Sarvazyan et al. 2011; Urban et al. 2011, 2012; Wells and Liang 2011; Aglyamov et al. 2012; Hansen et al. 2012; Konofagou et al. 2012; Litwiller et al. 2012; Sarvazyan and Egorov 2012; Sinkus et al. 2012).

CHAPTER 8

Elasticity, Nonlinearity and Viscoelasticity of Soft Tissues

8.1 INTRODUCTION

As discussed in previous chapters, it is of great interest for both academics and clinicians to study the mechanical properties of biological soft tissues for purposes including diagnosis and treatment outcome assessment. However, the mechanical properties of soft tissues are quite complex, which cannot be described simply using a one-parameter pure elasticity theory. In soft tissue elasticity measurement, typical phenomena of nonlinearity, viscoelasticity, anisotropy and heterogeneity are seen, which are introduced in the current chapter.

A typical cyclic mechanical test of biological soft tissues, for example indentation, can be seen in Figure 8.1, where several phenomena can be observed clearly. First, the loading and unloading curves for each cycle are different, called hysteresis. Second, the loading curve typically shows the shape of a bowl bottom of clear nonlinearity. Third, the loading–unloading curve of each cycle at the beginning of the test moves towards the right side of the x-axis and this curve becomes more and more consistent as the number of cycles increases. This phenomenon is called pre-conditioning. Because of the existence of pre-conditioning, a pre-test (several cycles of loading and unloading) is commonly adopted before collecting real data for analysis. Because of nonlinearity, a toe region with a very small stiffness

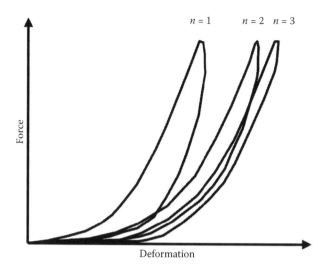

FIGURE 8.1 Typical force–deformation curve for the mechanical testing of soft tissues.

is always observable at a small value of deformation. Therefore, a pre-load is always necessary to reduce the effect of the toe region in practical measurement of soft tissue elasticity. A common phenomenon observed for the nonlinearity of soft tissue is the increase of tissue stiffness with the increase of strain level, which is called 'strain hardening'. A good example is the elasticity of cornea, which increases with the increase of intraocular pressure (IOP), thereby inducing a strain in the cornea. During mechanical testing, if the tissue behaviour significantly deviates from linearity, then nonlinearity should be considered.

In addition to the nonlinearity, another material property that should be considered is viscosity, as seen in the phenomena of hysteresis and pre-conditioning of the typical testing curve shown in Figure 8.1. Combined with the elastic properties, the term 'viscoelasticity' is commonly used to indicate the complex mechanical behaviours of soft tissues. Viscoelasticity can be easily demonstrated in a series of dedicated mechanical tests, including creep and force relaxation. Typical curves of creep and force relaxation tests are shown in Figure 8.2. In either test, if a step change of one variable (force or deformation) is induced, then the other variable (deformation or force) will change abruptly. After that, it will continuously change until reaching a final steady status. The phenomenon of viscosity is especially obvious if there is a large proportion

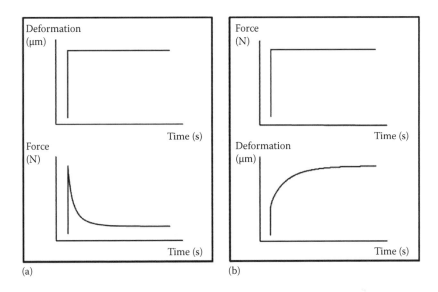

FIGURE 8.2   Typical curves of force relaxation (a) and creep tests (b).

of liquid in the tissue, such as in the pathology of oedema. Because of the fluid, the behaviour of mechanical test on soft tissues shows significant dependence on the testing speed, and therefore it is necessary to give details on the information of the test speed when reporting results of a strain–stress-related test. For shear wave propagation test, the frequency of the wave should always be given, as the measured propagation speed or derived modulus can be frequency-dependent. When the viscosity of the soft tissue is obvious, a viscoelastic model is necessary to calculate the separate elastic and viscous properties for an accurate characterisation of the soft tissue mechanics.

Although nonlinearity and viscoelasticity have long been recognised as intrinsic material properties associated with soft tissues, there is no consensus on how these properties can be reliably measured in practical testing conditions and used for clinical applications is still unclear. Recently, some groups have been trying to use either the nonlinearity or viscosity for tissue characterisation or potential use for diagnosis (Fatemi and Greenleaf 2002; Green et al. 2008; Oberai et al. 2009; Hall et al. 2011; Royston et al. 2011; Goenezen et al. 2012; Sack et al. 2013). However, efforts are still needed to generalise the results and make them practical for *in vivo* applications. In the following sections, we discuss how the nonlinearity and viscosity arise and how to model them in mechanical tests.

8.2 ORIGINS OF NONLINEARITY AND VISCOELASTICITY

The origin of the elasticity of a solid material can be traced to various types of atomic bonding (ionic, covalent, metallic, Van der Waals, etc.) depending on the material microstructure. In an equilibrium state of attractive and compulsive forces between the atoms, each atom in a material remains in a stable position. When mechanically stressed, it has the potential to go back to its original position, which gives the material the capability to resist an external force and shown as the elasticity of the material. Normally, it is feasible to assume a linear elasticity for the material when the strain is small within a certain range. However, when the strain becomes larger, the assumption of a linear elasticity can be easily violated. This phenomenon can be more easily observed in biological tissues such as collagen fibres, cells, water, etc., because of their complicated structure and composition. The critical value of strain below which linear elasticity can be assumed is different for different materials. For example, fatty tissue can still show a good behaviour of linear elasticity even with a large strain value of 30%; however, glandular tissue and, particularly, infiltrating ductal carcinoma increase quickly in stiffness with the increase of strain levels (Krouskop et al. 1998).

The viscosity of soft tissues mainly comes from their internal fluid. A very obvious demonstration of viscosity due to the change of fluid is seen in oedema. In oedematous tissue, the excessive accumulation of fluid in the interstitium makes the fluid hard to be exuded when indented, leaving an obvious pitting on the skin. The viscosity of soft tissues is determined by not only the intrinsic viscosity of the fluid itself but also its interaction with the solid part, mainly the friction. The contribution of the fluid to the tissue viscosity also depends on its location and existing format. Water makes up 60%–70% of the human body, which can be basically divided into two types: intracellular (two-thirds) and extracellular (one-third). With the advancement of new technologies, now it is possible to microscopically study the change of intracellular viscosity in physiology (Yanai et al. 1999; Kuimova et al. 2009). Generally, when there is pathological change of soft tissues, the internal fluid also has some variations, and therefore it is possible to diagnose the tissue condition based on the change of viscosity of the fluid.

Soft tissues have different microstructures. For example, in connective tissue the main components include fibres, cells and ground substances. Intrinsically, most of these tissue components are nonlinear and viscoelastic. Single collagen fibrils are viscoelastic, as demonstrated in creep and stress relaxation tests (Shen et al. 2011). However, the study of collagen fibres' mechanical properties also showed that the relaxation time of such fibres is relatively small compared to the tissue-level relaxation time, suggesting that other components such as proteoglycans are responsible for the tissue-level relaxation behaviour. Finding out which components of the tissue are mainly responsible for the nonlinear viscoelastic properties is an important issue for diagnosis when significant changes of the mechanical properties of the material are found.

8.3 MODELLING OF TISSUE NONLINEARITY AND VISCOELASTICITY

The modelling of nonlinear viscoelasticity of soft tissues is not an easy task, and classical mathematical representations of laws appropriate for the analysis have been reviewed (Drapaca et al. 2007; Wineman 2009). The mathematical analysis is complicated and, in order to use these theories directly, several assumptions such as homogeneity and isotropy need to be made. Even so, a direct analysis is commonly used in simple tests such as uniaxial indentation with simple test boundaries. In order not to make the explanation too complicated, here we introduce several simplified methods that have been commonly used in the literature for nonlinear viscoelasticity analysis.

8.3.1 Direct Stress–Strain Analysis

The simplest way to study the nonlinear elasticity of the soft tissues is to use the stress–strain or force–deformation relationship and to analyse the strain-dependent elasticity of the tissue. The first step is to calculate the Young's modulus in a small range of strain (e.g. 0–0.03) and regard it as the Young's modulus at the central point of the strain range. Then the elasticity of the soft tissue is calculated with the strain range shifted to a higher level (0.03–0.06) or with an increase of the strain range (0–0.06). The change of Young's modulus with the change of strain level can be used as an indication to represent the nonlinear elasticity of the tissue. Krouskop et al. (1998) studied the breast tissue using a direct analysis of the force–deformation relationship in an indentation test. The Young's modulus of normal and pathological breast tissues were measured under two different pre-compression strain levels of 5% and 20%. A different pattern of Young's modulus change with strain level was found among different types of tissues, such as fat, glandular tissue, fibrous tissue and ductal carcinoma tissue, which provided useful information for the diagnosis of breast pathology using elastography. Rome et al. used an exponential curve to describe the nonlinear relationship between the indentation force and deformation, and the two coefficients of the exponential function were used as quantitative parameters for comparison between different pathological groups (Rome and Webb 2000; Rome et al. 2001). The limitation of this direct analysis of the stress–strain relationship comes from the effect of viscosity. Because of viscosity, the elastic behaviour is dependent on the strain rate, and therefore for fair comparison, it is necessary to control the testing speed and clearly describe these testing conditions when reporting the results.

8.3.2 Lump Element Analysis

Another method for analyzing tissue viscoelasticity is to use a lump element analysis, similar to that used in the analysis of a circuit. Elastic elements – spring and viscous elements – dashpot are used to represent the contribution from elastic and viscous materials, respectively. A spring has pure elastic properties, which can be described by a linear relationship between the force and deformation:

$$F = kx \tag{8.1}$$

where k is the elastic constant. Compared to a spring, a dashpot is sensitive to the change of deformation, i.e. speed, in its mechanical behaviour:

$$F = \eta \dot{x} = \eta v \tag{8.2}$$

where
 η is the viscous coefficient
 v is the speed of the dashpot

Serial and parallel connections of these components can be used to simulate the complex behaviour of mechanical test: for example, the Maxwell, Voigt and Kelvin models as described in detail in Fung's classical book on biomechanics (Figure 8.3) (Fung 1993a). The basic steps to solve a practical problem include, first, embedding sufficient components in a proper mechanical model and then using experimental test data to find the best values for those components.

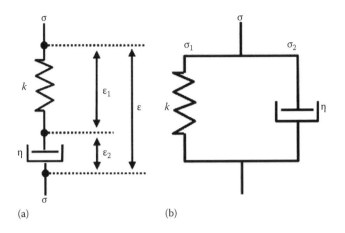

FIGURE 8.3   Elements of the viscoelastic models. (a) Maxwell model. (b) Kelvin model.

8.3.3 Quasilinear Viscoelastic (QLV) Model

Another popular model that can greatly simplify the analysis of viscoelasticity is the quasilinear viscoelastic (QLV) model initially proposed by Fung (1993a). The main idea is to separate the stress response of a small change of strain into an elastic part and a viscous part, as shown in a relaxation function (Fung 1993a):

$$K(\lambda, t) = G(t)T^{(e)}(\lambda) \tag{8.3}$$

where
λ is the instantaneous strain
t is the time with respect to the induction of λ
$G(t)$ is a reduced relaxation function
$T^{(e)}(\lambda)$ is an elastic function, which depends only on the induction of an instantaneous strain λ at time 0

According to the superposition principle, the stress response at time t can be extracted from the strain history λ(t):

$$T(t) = \int_{-\infty}^{t} G(t-\tau)T^{(e)}(d\lambda) = \int_{-\infty}^{t} G(t-\tau)\frac{\partial T^{(e)}(\lambda)}{\partial \lambda}\frac{\partial \lambda}{\partial \tau}d\tau \tag{8.4}$$

Through some manipulation by assuming the motion starting from $t > 0$, and $\partial T^{(e)}/\partial t$ and $\partial G/\partial t$ being continuous for $t > 0$, Equation 8.4 can be rewritten as

$$T(t) = T^{(e)}\left[\lambda(t)\right] + \int_0^t T^{(e)}\left[\lambda(t-\tau)\right]\frac{\partial G(\tau)}{\partial \tau}d\tau \qquad (8.5)$$

which is the typical form of a QLV model used to simulate the viscoelastic behaviour of soft tissues. This means that the stress at any time t is equal to the instantaneous stress response plus another part that is dependent on the past history of strain. As $G(t)$ is normally a decreasing function with a negative slope with respect to t, the stress is normally smaller than the instantaneous counterpart $T^{(e)}\left[\lambda(t)\right]$. This QLV model can also be written in a form of a creep behaviour for describing the strain using a history of stress. Readers can refer to Fung's classical biomechanics book for more details (Fung 1993a).

The next step is to decide the elastic function $T^{(e)}(\lambda)$ and the reduced relaxation function $G(t)$. $T^{(e)}(\lambda)$ can be obtained by a fast loading of the tissue, and the instantaneous response can be obtained immediately after the loading, because at this moment the effect of stress relaxation is small and can be neglected. The relaxation function $G(t)$ is commonly assumed to be in the form of a combination of a series of exponential functions:

$$G(t) = \frac{\displaystyle\sum_{i=1}^N C_i e^{-v_i t}}{\displaystyle\sum_{i=1}^N C_i} \qquad (8.6)$$

Normally, at least two components, a constant C_0 ($v_0 = 0$) and a non-zero exponential coefficient C_1 ($v_1 \neq 0$), are used to describe the relaxation behaviour, with $C_0 + C_1 = 1$ at time 0 and C_0 representing the final steady status when t is infinite.

Alternatively, these elastic and viscous parameters can be obtained through an experimental fitting process, where the strain history can be used as an input of the simulation. In this case, the nonlinear elasticity for biological tissues can be simply modelled using the hyperelastic theory by defining a strain energy density function (Horgan and Saccomandi

2002). Readers who are interested in these hyperelastic models can refer to some excellent work on this topic (Gasser et al. 2006; Martins et al. 2006; O'Hagan and Samani 2009).

The simulated stress response can be compared with the measured experimental data, and an objective function can be defined to optimise those model parameters as the tissue properties, which can be done using a simple numerical simulation analysis (Zheng and Mak 1999; Huang et al. 2005) or a more complex finite element analysis (Tonuk and Silver-Thorn 2003; O'Hagan and Samani 2009). The optimised material parameters can be used to study the differences of material properties between different pathologies (Huang et al. 2005; O'Hagan and Samani 2009).

8.4 METHODS FOR MEASUREMENT OF VISCOELASTICITY *IN VIVO*

The creep and force relaxation tests described in Section 8.1 are mainly used for *in vitro* tests for quantifying biological tissue properties, and they are not suitable for *in vivo* tests, as the boundary conditions of a tissue cannot be well controlled. For the *in vivo* test, the measurement of viscoelastic properties is a difficult task. Normally, a simplified model incorporating a few elastic and viscous parameters is proposed to account for the complex behaviour of the tissue. With this simplified mechanical model, a theoretical solution may be obtained for the specific test conducted on the tissue, e.g. the shear wave dispersion ultrasound vibrometry (SDUV) (Urban et al. 2012). Refer to Equation 6.11 for the details of this method. It can be easily seen that the shear wave speed changes with the vibration frequency. Therefore, in this case, multiple frequency data can be collected and regressed to obtain both the elastic and viscous properties of the tissue. On the other hand, if theoretical solutions are not possible, then finite element analysis will serve as a computational method to solve this problem. For each element in FEM, a specific viscoelastic model can be used and then an inverse method can be adopted to extract the viscoelastic parameters of each element in the model (Samur et al. 2007).

8.5 SUMMARY

Although nonlinear elasticity and viscosity are distinct material properties that can be used for the differentiation between normal and pathological tissues, the applications are scattered in the literature and no pathway to date has been paved guiding such applications in soft tissue characterisation. This field is still in its infancy and more efforts are needed. In future studies, the combination of elastic properties, nonlinear and viscous parameters can be used for a better characterisation of soft tissues, which will provide important information for the assessment or diagnosis of their pathological conditions.

CHAPTER 9

Finite Element Methods and Inverse Solutions for Elasticity Measurement

9.1 INTRODUCTION

The measurement and imaging of elasticity aim to estimate the mechanical properties, including, but not limited to, Young's modulus. This chapter documents the efforts of several groups on studying the mechanical properties of biological tissues by using modelling techniques.

Elasticity modelling should start from the formulation of a real measurement or imaging problem, which is commonly expressed by partial differential equations (PDEs). If static stimuli are applied, the source term in PDEs can be simplified to zero while the vibrational stimuli induce a time-dependent term. Such a mathematical simplification for the static stimuli may not be an advantage in practice, since the PDEs have to be determined entirely by the usually poorly known boundary conditions due to the complexity of biological tissues and their surroundings (Barbone and Oberai 2010).

Analytical methods can be employed to solve the PDEs with simple geometries and boundary conditions. However, numerical methods are more powerful to solve the equations on irregular domains, even for an elastically heterogeneous tissue. For this purpose, both the finite difference method and the finite element method (FEM) can be applied. However, there is no doubt that the FEM, with many well-established commercial packages, is the most popular approach for solving PDEs. The whole process involves the following steps (Doyley 2012):

- Segment the geometry of tissue into a series of finite elements by mesh generation.

- Derive a weak form of the PDEs using either the variational or weighted residual method.

- Substitute a basis or shape function to produce a system of linear algebraic equations.

- Impose the external boundary conditions to solve the resulting equations.

In tissue elasticity modelling, one of the main directions is the realisation of elastography within the framework of solving an inverse problem, given the measured displacement fields, an assumed form of the tissue's constitutive equation and the law of conservation of momentum (Barbone and Oberai 2010). Rather than an approximate estimate of elasticity, they provide more reliable estimates of shear modulus and other mechanical parameters, namely viscosity, anisotropy, poroelasticity and nonlinearity (Doyley 2012). Some efforts have also been made on the safety assessment of new techniques, especially those using the acoustic radiation force (Castaneda et al. 2013).

In the following sections, we summarise the physics foundation for the current approaches to estimating mechanical parameters, which generates the governing PDEs that describe the core problem of elasticity measurement and imaging. To simplify our discussion, these approaches are classified into two groups by using the quasi-static stimuli and the time-dependent stimuli, respectively. After that, we provide a brief review of the results achieved in the field of elasticity modelling.

9.2 PHYSICS FOUNDATION

Soft tissues exhibit linear elastic behaviour under infinitesimal deformation, which relates the strain tensor ε_{kl} to the stress tensor σ_{ij} as follows (Fung 1993d):

$$\sigma_{ij} = C_{ijkl}\varepsilon_{kl} \tag{9.1}$$

where the tensor C is the Christoffel rank-four tensor consisting of 21 independent elastic constants. Assumptions applied to soft tissues can reduce the number of independent elastic constants. If the soft tissue is assumed to be purely elastic and isotropic, its mechanical behaviour can be described by only two independent constants: the Lamé constants λ and μ (or the shear modulus). They are related to Young's modulus E and Poisson's ratio ν by (Raghavan and Yagle 1994)

$$\lambda = \frac{E\nu}{(1+\nu)(1-2\nu)} \tag{9.2}$$

$$\mu = \frac{E}{2(1+\nu)} \tag{9.3}$$

For nearly incompressible soft tissues, ν is close to 0.5, so μ is one-third of Young's modulus, and λ converges to a certain value. Equation 9.1 can be further simplified to describe such incompressible isotropic linear elastic soft tissue by

$$\sigma = E\varepsilon \tag{9.4}$$

Since stress distribution cannot be measured *in vivo*, conventional quasi-static elastography (Ophir et al. 1991) was developed with the assumption of a uniform stress distribution. It makes tissue strain a qualitative measure, and leads to strain image artefacts in practice. Some studies estimated Young's modulus with prior knowledge on boundary conditions and developed the indentation techniques, but these techniques only provide the one-dimensional measure (Hayes et al. 1972). Equation 9.4 describes only the problem of quasi-static elastography. Now we want to establish the common ground for measuring systems irrespective of whether one uses the quasi-static force or the time-dependent stimuli (Doyley 2012). The direct elasticity problem can be first described in a compact form from the conservation of linear momentum (Fung 1993d):

$$\nabla \cdot \left[\sigma_{ij}\right] = f_i \tag{9.5}$$

where
　∇ is the del operator
　σ_{ij} is the stress tensor
　f_i is the deforming force

According to Newton's second law of motion, here f is proportional to $\ddot{u}$ (i.e. the second derivative of displacement vector u). Since stress cannot be measured *in vivo*, it is typically eliminated by using the following relationship for linear, isotropic elastic materials:

$$\sigma_{ij} = 2\mu\varepsilon_{ij} + \lambda\delta_{ij}\Delta \tag{9.6}$$

where δ_{ij} is the Kronecker delta, $\Delta = \nabla \cdot u = \varepsilon_{11} + \varepsilon_{22} + \varepsilon_{33}$ relates to compressibility and the components of the strain tensor are defined as $\varepsilon_{ij} = \dfrac{1}{2}\left(\dfrac{\partial u_i}{\partial x_j} + \dfrac{\partial u_j}{\partial x_i}\right)$. The resulting equations of equilibrium called the Navier–Stokes equations are given by (Parker et al. 2011)

$$\nabla \cdot \mu\nabla u + \nabla(\lambda + \mu)\nabla \cdot u = \rho\ddot{u} \tag{9.7}$$

where ρ is the density of the material. Actually, with given boundary and initial conditions, the equations do not lose the generality for a viscoelastic material if the two Lame constants λ and μ are complex.

Equation 9.7 governs the response of a homogeneous, isotropic, linearly elastic material to stimuli. When loads are applied quasi-statically, the right-hand side is negligible and can be set to zero. Further, considering the soft tissue as an incompressible material, the dilatation $\nabla \cdot u$ is equal to zero, and the equation is simplified to

$$\nabla^2 u = 0 \tag{9.8}$$

Because of the usually unknown boundary conditions, the inverse problem will be poorly conditioned, which affects the measurement and imaging quality.

On the other hand, when loads are harmonic or transient, the dynamic response is described in terms of waves propagating within the tissue. Shear waves and pressure waves propagate independently in the material, and they interact only at the boundaries. The shear wave will not induce a volume change within the material, so the dilatation $\nabla \cdot u$ is equal to zero, and the shear wave equation derived from Equation 9.7 can be written as

$$\nabla^2 u = \frac{1}{c_s^2}\ddot{u} \tag{9.9}$$

where $c_s = \sqrt{\mu/\rho}$ is the shear wave speed. This equation can either be solved in terms of standing waves or propagating waves depending on the particular conditions. It is the physics basic for most ultrasound-based elasticity imaging, including sonoelastography (Gao et al. 1995), crawling wave sonoelastography (Wu et al. 2004b), supersonic shear imaging (SSI) (Bercoff et al. 2004a) and spatially modulated ultrasound radiation force (SMURF) imaging (McAleavey and Menon 2007).

9.3 MODELLING AND SIMULATION

The Navier–Stokes equations (9.7) are the governing PDEs for the elasticity problem. Both analytical and numerical methods can be utilised to solve forward and inverse problems. The former aims to find the displacement $\boldsymbol{u}$ with the known mechanical parameters, such as shear modulus μ, while the latter implements a mapping from the measured displacement to the mechanical parameters. Under the concept of inverse problems, much effort has been made to improve the reconstruction of mechanical properties by using modelling techniques. Besides, with the advent of acoustic radiation force, which is a phenomenon associated with the force induced by the propagation of acoustic waves, some studies have focused on the feasibility and safety of the so-called remote palpation techniques. In this section, we will present the achieved results by using modelling techniques. We will classify the literature into four research directions: inverse problem on quasi-static elastography, modelling related to wave propagations, consideration of acoustic radiation force and modelling indentation tests.

9.3.1 Inverse Problem on Quasi-Static Elastography

Conventional quasi-static elastography assumes a uniform stress distribution, and thus interprets the internal strain as a measure of shear modulus. However, this assumption is not well established, and the spatial derivative will amplify the high-frequency noise of displacement measurement (Castaneda et al. 2013). To overcome these drawbacks, Kallel and Bertrand (1996) treated the reconstruction of elastic properties from the axial component of the 3-D displacement vectors as a nonlinear least-squares problem. Several groups extended the reconstruction approach using two or three components of the displacement vectors and also modified the solver for the nonlinear problem (Li et al. 2008a; Richards et al. 2009). In a recent study, Eder et al. (2012) reformulated the inverse reconstruction problem as an optimisation problem and proposed the trust region reflective

Newton solver to achieve a convergence of the reconstruction process. In the study of Pavan and Carneiro (2011), the stress–strain data were first simulated using the Vernda–Westmann constitutive hyperelastic model, and then the strain contrast between the inclusion and background was compared for materials with nonlinear strain–stress behaviour. The modelling results showed that material nonlinearity as well as small strain Young's modulus had an influence on the strain image contrast.

9.3.2 Modelling Wave Propagation

The Navier–Stokes equations describe the behaviour of a shear wave within a homogeneous medium thicker than the shear wavelength. However, thin-layered tissue structures such as the skin and cornea will guide the propagation of shear wave, and this leads to dispersive effects. Nguyen et al. (2011) validated shear wave spectroscopy as a method for Young's modulus quantification in such layered tissues. In their work, shear wave dispersion curves in thin layers were obtained by finite-difference simulations as the numerical solution of the boundary conditions. The simulation results were consistent with the analytical formula fitted to the experimental data (Couade et al. 2010) and the theoretical solution deduced from the leaky Lamb wave theory. Further simulation results emphasised the fact that the longitudinal component leaks into the surrounding liquid, whereas the transverse component remains confined within the plate. Besides, unlike the propagation in a bulk medium, the shear wave velocity of the non-viscous plate presented dispersive effects because of the guided propagation, but no significant differences were found on the dispersion curves between the viscous and the non-viscous plate.

Modelling techniques have also been applied to crawling wave sono-elastography. McLaughlin et al. used numerical methods to calculate the speed of the moving interference pattern, which is normally about 1/300 of the shear wave velocity (Wu et al. 2004b; McLaughlin et al. 2007). In the further work of Hoyt et al., simulation results validated a viscoelastic approach based on sonoelastography imaging (Hoyt et al. 2008b).

9.3.3 Consideration of Acoustic Radiation Force

Acoustic radiation force (ARF) gives rise to the so-called remote palpation, which pushes the tissue internally to examine the local elasticity or generate vibration to propagate inside the tissue. This force is caused by the energy density gradient that occurs in the medium, arising either from absorption or reflection of the wave (Nightingale et al. 1999). The earliest imaging attempt with ARF may be acoustic radiation force impulse (ARFI) imaging (Nightingale et al. 2002). The biggest concern of ARFI is the safety problem induced by the high-intensity push beams. Palmeri and Nightingale (2004) evaluated the effect of temperature increase under several pushing schemes in ARFI. They assumed that the change of sound speed with respect to temperature is linear for temperature changes less than 6°C. FEM simulation results revealed the heating associated with ARFI imaging. They concluded that temperature increase is greater for less absorbing tissues, and ARFI images may suffer from artefacts due to sound speed changes.

Different studies have investigated better schemes to reduce the ultrasound intensity. Bouchard et al. (2009a) evaluated schemes that were proposed to reduce the acquisition time, and heating was evaluated through FEM modelling and thermocouple measurements. It was concluded that the frame acquisition time can be reduced and heating issues mitigated when a new beam acquisition sequence is employed with a parallel receive beam-forming. Nii et al. (2013) assumed that the amplitude of tissue displacement was caused by ARF via mode conversion, so they proposed a new strategy in which only the tumour regions confirmed in advance were targeted for elasticity measurement using ARF. They investigated the effect of frequency and incident angle of a longitudinal wave on generating the shear wave. They confirmed that a shear wave could easily be generated at an elasticity boundary where the difference of elasticity is large and mode conversion efficiency depended on the incident angle of a transmitted longitudinal wave. However, the amplitude of the displacement generated by the shear wave was only 1/10 of the ARFI, which may not be large enough for imaging.

Besides, FEM modelling has been applied to evaluate a 1.5-D ARF transducer for ARFI by Dhanaliwala et al. (2012). When compared to a conventional 1-D array, simulation results showed that the new transducer reduced echo decorrelation and significantly improved displacement estimation in terms of the signal-to-noise ratio (SNR) and the contrast-to-noise ratio (CNR).

9.3.4 Modelling Indentation

As described in Chapters 3 and 4, analysis of indentation tests can be very complicated when the boundary conditions are complicated. Therefore, FEA is a commonly used method for the analysis of the indentation (Figures 9.1 and 9.2) as well as suction tests (Hendriks et al. 2004;

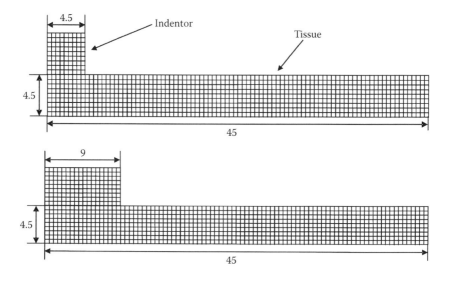

FIGURE 9.1 Two FEA mesh models (half cross section) for indentation tests with different indenter radii of 4.5 and 9.0.

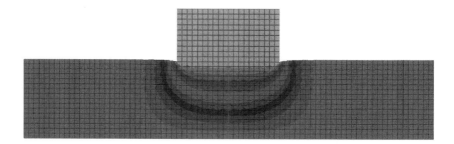

FIGURE 9.2 Finite element modelling showing strain distribution under an indentation test.

Mazza et al. 2008). FEA has been used to investigate the finite deformation effects of indentation and provide compensation factors for large deformation (Zhang et al. 1997). Lu and Zheng (2004) used FEA to study the effects of curved substrates on the indentation results (Figure 9.3). They established a series of FE models with different ratios between the indenter radius and tissue thickness, as well as different ratios between indenter radius and substrate curvature, mimicking different bones. In addition, FEA has also been widely used for the analysis of suction test (Barbarino et al. 2011). If the time taken for establishing an FE model and calculation

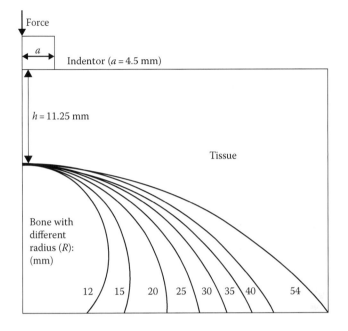

FIGURE 9.3 Series of indentation models with a curved substrate embedded in the tissue. The radius of the bone ranged from 54 to 12 mm, i.e. a/R ranged from 0.083 to 0.375 (a/h = 0.4, a = 4.5 mm, h = 11.25 mm), where a is the radius of the indenter, h is the thickness of the test tissue and R is the radius of the substrate bone.

can be reduced, FEA can be used to extract more intrinsic and accurate biomechanical properties from indentation and suction tests in the future.

9.4 CONCLUSION AND FUTURE RESEARCH DIRECTIONS

In this chapter, we provide an overview of the efforts made on modelling for elasticity measurement and imaging. We first explain the physics foundation that formulates the elasticity modelling problem for most approaches to estimate mechanical parameters of soft tissue. Then we review some modelling results and briefly state their findings. As the focus is on the reconstruction of shear modulus, some advanced topics of elasticity are not fully included, such as viscoelasticity, poroelasticity, nonlinearity and anisotropy. Although these parameters may significantly increase the difficulty of modelling and reconstruction, they will allow a better approximation of soft tissue behaviour under test. For example, several groups have observed that muscles and tendons have V-shaped shear wave patterns. It is unlike the concentric pattern that appears in an isotropic material, which may be ascribed to anisotropy (Uffmann et al. 2004), but hardly any work has been devoted to solving it (Sinkus et al. 2005). Actually, we still have a long way to go to fully understand and to accurately measure the mechanical characteristics of soft tissue. FEA will be a very powerful tool in the further development of new methods for tissue elasticity measurement and imaging.

CHAPTER 10

Clinical Applications of Soft Tissue Elasticity Measurement

10.1 INTRODUCTION

Measurement of tissue elasticity is important for understanding the physiology and pathology of a tissue, as elasticity or viscoelasticity is naturally associated with any human tissue structure, from the levels of molecule, cell, tissue, up to organ. Better understanding of tissue elasticity can help us investigate tissue development, growth, degeneration, regeneration, malfunction, cancer formation and so on. Tissue elasticity information can be used for screening, diagnosis, progression monitoring, treatment planning and outcome measurement. It is also essential for surgery training using virtual reality. The biomechanical parameters of tissues are also crucial for modelling tissues and organs, such as using finite element analysis. Without accurate inputs of tissue elasticity, it is not possible to obtain correct simulation results. In this chapter, we summarise some typical applications of tissue elasticity measurement, categorised into following subsections according to the tissue types. We exclude the introduction in the application of cell elasticity, which is closely linked with mechanobiology and is a very hot topic currently. Interested readers can refer to books and publications related to mechanobiology (Wang and Thampatty 2006; Suresh 2007).

10.2 APPLICATIONS IN UNDERSTANDING TISSUE PHYSIOLOGY AND COMPUTER MODELLING

Elasticity is important for maintaining the normal physiological functions of tissues, especially load-bearing tissues, such as the articular cartilage and musculoskeletal tissues. Macroscopically, the tissue components are maintained in an equilibrated metabolic state at an appropriate level of new cell growth and apoptosis so that the mechanical properties of tissues are optimised for performing their normal physiological functions. Microscopically, it has been demonstrated that the biomechanical microenvironment is very essential for the stem cell lineage specification (Engler et al. 2006). Soft extracellular matrix (ECM) leads to neurogenic differentiation. ECM with moderate elasticity is advantageous for myogenic differentiation, while a stiff ECM is osteogenic. Special aspects of elasticity related to physiology include the normal ageing effect, diurnal or menstrual change and variation of elasticity associated with muscle contraction. For example, skin elasticity is known to change with age, and the elderly usually have stiffer skin than young people. The suction method has been used to study the effect of ageing on the skin's mechanical properties (Ohshima et al. 2013). On the other hand, short-term physiological changes due to hormonal fluctuation such as the menstrual cycle have also been demonstrated to induce a change of elasticity to soft tissues such as the breast parenchyma (Lorenzen et al. 2003). The variation of breast elasticity in different quadrants and also the relationship between the breast elasticity and other factors such as body mass index (BMI) and breast size were investigated in another study, which is useful for a better understanding of the breast physiology (Li 2009). Circadian change of the tissue stiffness has also been investigated, although it failed to demonstrate a significant difference in arterial tissues (Kollias et al. 2009; Drager et al. 2011). Further study is needed to investigate how different parts of the human body such as lower and upper extremities will be affected diurnally in terms of elasticity change.

10.2.1 Skeletal Muscles

Skeletal muscles generate force and provide support and locomotion for the whole skeletal system. Muscles are optimised by their morphological and biomechanical properties to realise their functions. It is known that the elasticity of muscular tissues changes significantly during contraction. Therefore, elasticity provides important information to study the basic physiology of the muscle status.

There are different definitions in the literature of muscle stiffness, which might be due to the difficulty in measuring the intrinsic stiffness of this tissue from the beginning. A common definition of muscle stiffness is the force generated by the muscle compared to change in length during contraction (Cook and McDonagh 1996), which can be called functional stiffness here. However, it is difficult to obtain the force generated by a single muscle, and muscle lengthening is not necessarily involved in all contraction types. Therefore, this method is not so easy to apply in clinical situations.

Indentation devices have been developed for the quantitative measurement of muscle stiffness. For example, the tissue compliance meter (TCM) is a device with a large disc-shaped ring attached at the end of a shaft where a rubber tip is used as the indenter (Fischer 1987b). The force applied is measured by a force gauge, and the indentation depth is recorded by the movement of the large ring along the shaft. The reading of the depth of indentation at a certain fixed force is recorded as a quantitative parameter for quantifying the tissue stiffness. This device was introduced for muscle studies such as the stability of paraspinal muscles (Sanders and Lawson 1992). In order to improve the operation reliability and measurement accuracy (Kawchuk and Herzog 1995), a new version of the TCM, the myotonometer, was developed with automatic measurement of force and indentation depth (Leonard and Mikhailenok 2000; Leonard et al. 2001). On the other hand, the tissue ultrasound palpation system (TUPS) is also a portable device with automatic recording of indentation data for the study of the biomechanical properties of the muscular tissues in the lower extremity. A four-parameter quasilinear viscoelastic study showed that different nonlinear viscoelastic parameters vary among subjects, sites and posture (Zheng and Mak 1999). The soft tissue layer including the muscle layer in the neck can be assessed in a quantitative way using the TUPS, which can be used for the objective assessment of fibrosis after radiotherapy (Zheng et al. 2000b; Leung et al. 2002). Qualitative compression elastography can also be used to study muscle elasticity. The general principle is to compare the strain ratio between different tissue components and obtain some qualitative or semi-quantitative information about the stiffness contrast for different compositions, such as between fascia and muscle (Cespedes et al. 1993). Recently, vibration-based shear wave elastography (SWE) techniques are becoming more and more popular to study muscle stiffness because it can give a quantitative assessment (Brandenburg et al. 2014; Klauser et al. 2014).

The change of muscle stiffness during a normal contraction process is of particular interest to physiologists because it may reflect the functional wellness of the muscles. An *in vivo* contraction experiment has demonstrated that muscle stiffness increases with the level of strength generated in isometric contraction (Nordez and Hug 2010; Shinohara et al. 2010). Compared to electromyography (EMG), it was found that the muscle stiffness provides a better prediction of the force generated by a single muscle than surface EMG (Bouillard et al. 2011). However, because of technical limitations, a conventional SWE system cannot measure the muscle stiffness when the activity level is high with respect to the maximum isometric voluntary contraction (MVC). Wang et al. (2014) developed a vibro-ultrasound shear wave velocity measurement system that utilised two A-lines in a linear array transducer to track the propagation of shear waves (Figure 10.1a). Because of ultra-fast acquisition capability of the ultrasound system, the velocity measurement range was significantly increased, so the whole range of MVC could be studied with respect to the muscle stiffness (Figure 10.1b).

10.2.2 Surgical Simulation

Computer simulation is another field where measurement of soft tissue elasticity can be of significant importance. Computer simulations of surgical procedures are becoming more and more popular because of the fast advancements in computer technology, including both hardware and software support. Through effective and real-time operation simulation, medical surgery training can become more accessible, flexible, efficient and ethical compared to traditional simulations using animals or cadavers (Misra et al. 2008). In the simulation, it is essential to provide not only a graphic real-time view of the operation environment but also a haptic feedback of the tool–tissue interaction to mimic a realistic situation. In order to accurately simulate the interaction between the surgical tool and tissue, a mechanical model is necessary, and feasible material mechanical properties are required to be included into the model. A simple linear elastic model is found to be fast and effective for simulating a small deformation, but it fails when the deformation is large (Kerdok et al. 2003). For simulating larger strains, it is necessary to incorporate nonlinear and viscoelastic tissue properties in the model (Miller 2000; Miller and Chinzei 2002). How to reliably obtain those mechanical properties for *in vivo* soft tissues in order to accurately simulate the tissue behaviour in a surgical operation is still an active research area (Lister et al. 2011; Shen et al. 2012). With the

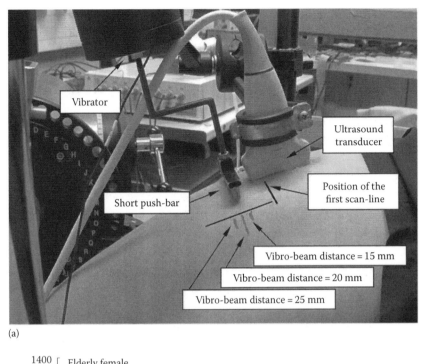

(a)

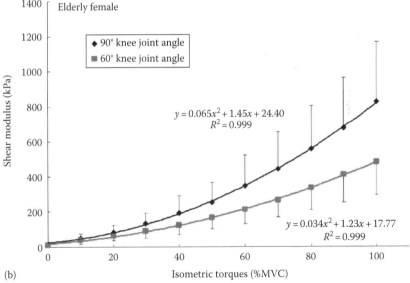

(b)

FIGURE 10.1 (a) Vibro-ultrasound system for the measurement of shear wave speed by two A-lines in an array transducer under isometric contraction. (b) Measured shear modulus change with isometric contraction level in young and elderly female groups.

advancement of various elasticity measurement techniques, tissue mechanical properties can be measured more easily, reliably and accurately, and the performance of surgical simulations which will be beneficial for a number of orthopaedic applications, such as training of surgical procedures for new doctors and pre-surgery planning, can thereby be improved.

10.3 APPLICATIONS IN DISEASE DIAGNOSIS AND TREATMENT MONITORING

To date, most applications of soft tissue elasticity measurement have been related to disease diagnosis and treatment monitoring. This is because soft tissues will usually undergo quite a distinct change of their constituents, possibly before the macroscopic symptoms of a disease appear, and then these pathological changes will obviously affect its mechanical properties. Soft tissue elasticity can either increase or decrease in a pathological condition compared to their normal status, depending on how the tissue changes in the course of the disease. When the tissue undergoes excessive proliferation of certain components such as collagen fibres, the density of the tissue normally increases and the stiffness increases correspondingly. In contrast, when the tissue undergoes some degenerative changes, such as breakdown of collagen fibre and loss of proteoglycans, the stiffness decreases, as seen in the degenerative change of articular cartilage in osteoarthritis. Palpation has been the oldest method for the purpose of elasticity assessment because it is easy and convenient. However, the hand palpation method has been criticised for its subjective and qualitative nature, and its heavy dependence on the experience of the operator hinders its widespread use in clinical practice. With the advent of advanced sensor technologies, especially in the last two decades, various tools have been developed for the quantitative and objective measurement of tissue elasticity, such as novel indentation methods, suction, compression elastography and SWE, as introduced in the previous chapters with and without inverse problem solution. Applications of soft tissue elasticity in this field are many, and a number of reviews have already been published (Mariappan et al. 2010; Parker et al. 2011; Sun et al. 2011; Wells and Liang 2011). In the following, we briefly review the applications of elasticity measurement to the specific diagnostic fields of liver fibrosis, cancer and tissue degeneration.

10.3.1 Liver Fibrosis

Liver fibrosis is a chronic disease of the liver, where excessive ECM is deposited and accumulated as an effort of wound-healing to repair liver damage. In the worst scenario, it will finally develop into the cirrhosis stage, which is one of the leading causes of death in Western countries (Friedman 2003; Bosetti et al. 2007). Traditional methods for the diagnosis of liver fibrosis mainly include liver biopsy and serum test. However, because of the limitations of these clinical methods, the need for development of a noninvasive technique for diagnosing or screening liver fibrosis still exists. Change of liver elasticity is an obvious feature associated with liver fibrosis, so various elasticity measurement techniques naturally have found applications in assessing this chronic disease.

Compression and indentation tests have been conducted on fresh liver samples *ex vivo* in order to verify the usefulness of elasticity measurement for liver disease diagnosis (Yeh et al. 2002). It was demonstrated that liver stiffness increases significantly with the increase of the fibrosis severity. Also, the nonlinearity of the liver stiffness with respect to the strain level is obvious (Yeh et al. 2002). Because of this, liver stiffness measurement *in vivo* has become an active research area since transient elastography (TE) became available for this purpose (Sandrin et al. 2003). Since then, a large number of papers (~2000 up to 2015) have been published focusing on comparison with traditional methods and demonstrating the clinical potential of this technique for the diagnosis of chronic liver diseases with different aetiologies and monitoring of treatment effects such as liver transplant (Castera et al. 2005, 2008; Ziol et al. 2005; Foucher et al. 2006 Friedrich-Rust et al. 2008). Generally, TE is regarded as an accurate and reliable method for the assessment of liver fibrosis; however, situations like the effect of meals, active hepatitis and obesity need special attention because the final results of liver stiffness measurement might be mixed by the effects of these factors or the probe type. Another difficulty in using the TE system is that there is no guide for the operation of the probe placed on the surface of the abdomen. Because blood vessels or bile ducts might be mistakenly selected in the measurement, which will significantly affect the measurement result, it is advantageous to have a graphic view where the measurement line is drawn for the measurement. Mak et al. (2013) developed such a B-mode imaging-guided TE system. Figure 10.2 shows the interface where the B-mode image is used as a guide to select the region of interest for measurement. Related validation test and the effect of different test protocols were also studied. This system has the potential of decreasing the uncertainty in the localisation step before real measurement and also increasing the user friendliness of the operation of the TE system.

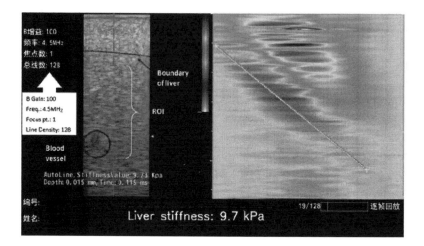

FIGURE 10.2 Transient elastography system Liverscan with guidance from ultrasound B-mode for the measurement of liver stiffness.

Magnetic resonance elastography (MRE) is another popular method that can be used for the measurement of liver stiffness (Muthupillai et al. 1995; Rouviere et al. 2006; Venkatesh and Ehman 2014). Specific design is necessary for inducing the shear wave propagating into the liver without affecting the normal operation of the MRI machine (Tse et al. 2009). MRE measurement of liver stiffness has the advantages of noninvasiveness and high measurement and diagnostic accuracy (Ichikawa et al. 2015). However, because of its high cost and limited accessibility, it is uncertain whether it can become a clinical routine procedure.

After some fundamental studies on the relationship between liver stiffness and pathologies to demonstrate its accuracy and reliability, nowadays liver stiffness measurement has become almost a clinical tool for the diagnosis of liver diseases, especially fibrosis. Different cutoff values have been suggested for the detection of different stages of liver fibrosis of different aetiologies (Castera 2011). The sensitivity and specificity for the diagnosis of liver fibrosis using the liver stiffness value has also been studied for both TE and MRE, which were found very promising for clinical studies (Ichikawa et al. 2015). Furthermore, the liver stiffness can also be used as a surrogate marker for the monitoring of treatment efficacies (Wong et al. 2011; Corpechot et al. 2012). It has been suggested that the progressive increase of the liver stiffness is an indicator of poor outcome for the treatment of primary bile cirrhosis (Corpechot et al. 2012). With more and more clinical data becoming available and standardisation of the measurement procedure, liver stiffness measurement would become an indispensable tool for the management of various chronic liver diseases.

10.3.2 Cancer

Cancer is another major field where the measurement of soft tissue elasticity may become a routine clinical tool for diagnosis and treatment assessment. Cancer is a condition of unregulated, excessive cell growth and may invade other tissues through metastasis in a later stage. The excessive tissue deposited in the tumour location associated with the angiogenesis makes the affected mass significantly different from the surrounding tissues in both density and mechanical properties. A well-known example is breast cancer, which was also the earliest target of applications when the technique of elastography was developed. Before elastography was applied to this field, a common clinical question was whether the tumour is benign or malignant when it is detected in mammography or ultrasound images. Such tumours can be seen commonly in these clinical imaging methods. However, it is uncertain which types these tumours belong to. Through an indentation test *in vitro*, it has been demonstrated that the malignant breast tumour is the stiffest part among the different components of breast tissues (Krouskop et al. 1998; Samani et al. 2007). Quantitative measurement of breast elasticity through ultrasound-based SWE also showed that the breast elasticity increases from normal to benign and from benign to malignant tumours. Using a commercial supersonic shear imaging (SSI) system, Chang et al. (2011) found that the average stiffness of a benign lesion was about 46 ± 43 kPa while that of a malignant lesion was 153 ± 58 kPa, which was significantly higher. Their results are quite consistent with a previous study performed using an SSI prototype (Tanter et al. 2008; Athanasiou et al. 2010). When combined with conventional B-mode imaging, the diagnostic performance can be achieved with an optimal level compared to their single use. MRE also has the potential for the assessment of breast lesions (McKnight et al. 2002; Xydeas et al. 2005), but a large-scale clinical study is still lacking in this field. It is believed that the clinical use of these elasticity measurement techniques, as a complementary method to traditional diagnostic methods, will reduce unnecessary biopsies for a large population of patients with benign lesions (Balleyguier et al. 2013).

In addition to breast cancer, other cancerous tissues can also be subjected to soft tissue elasticity studies for the diagnosis and treatment effect assessment, the most potential areas being brain, prostate and skin cancers (Di Ieva et al. 2010; Liang et al. 2010; Good et al. 2014). For brain tissue elasticity measurement, currently MRE is the only choice because both ultrasound and light cannot penetrate the skull. The consistency between MRE measurement and the stiffness assessed by neurosurgeons has been established (Xu et al. 2007), and the change of brain tumour elasticity compared with that of normal white matter is found to be different for different tumours (Xu et al. 2007; Simon et al. 2013). A malignant brain tumor can be significantly softer than the normal brain tissue (Simon et al. 2013). For prostate, the elasticity measurement/imaging can be achieved through transrectal ultrasound imaging. In addition to diagnosis, one of the main applications of elasticity measurement/imaging is to guide biopsy (Pallwein et al. 2007; Good et al. 2014). Skin cancer is one potential area where optical imaging elastography can find broad applications because of its high spatial resolution (Liang et al. 2010).

10.3.3 Tissue Degeneration

Tissue degeneration also comprises a large proportion of the various diseases. For example, it can happen to various types of tissue, such as Alzheimer's disease in neural tissues, amyotrophic lateral sclerosis in muscular tissues and cartilage and disc degeneration in connective tissues. Here we focus on articular cartilage degeneration, which is heavily involved in the pathology of osteoarthritis. Articular cartilage is a thin tissue covering the bony ends, providing the functions of lubrication and cushioning during joint movement. The early detection of articular cartilage degeneration is especially important for the management of osteoarthritis, because currently there is no cure when the cartilage degeneration has reached the late stage. Clinically, doctors use a blunt probe inserted through an arthroscopic channel to palpate the stiffness of articular cartilage, which heavily depends on the operator's experience and is qualitative, rendering it difficult for comparison among operators. Historically, compression, confined compression and indentation have been the most commonly used methods for measuring the cartilage stiffness *in vitro* (Hayes et al. 1972; Mow et al. 1980). Only indentation is appropriate to be applied for the mechanical test of living tissues. An arthroscopic indention device, later combined with an ultrasound transducer at the tip, was developed for the assessment of articular cartilage (Lyyra et al. 1995; Laasanen et al. 2002). Automatic indentation can also be achieved by incorporating a robotic arm or motor-driven fast indentation (Niederauer et al. 2004; Samur et al. 2007). Although a correlation between the cartilage stiffness and its degeneration level, such as the Mankin score or International Cartilage Repair Society

(ICRS) score, has clearly been demonstrated for *in vitro* cartilage samples, data for *in vivo* measurement of cartilage stiffness are still scarce (Lyyra et al. 1999). An accurate and reliable technique that can be used for measuring the mechanical properties of cartilage in a convenient and fast way will be beneficial for the development of new drugs or treatment schemes for osteoarthritis management. Another similar situation is the intervertebral disc, which is an important tissue for supporting the spine movement and assumed to be related to many pathologies such as neck pain and low back pain. Similar to cartilage, the disc is associated with a decrease in its mechanical properties after its degeneration (Iatridis et al. 1998; Cortes et al. 2014). Recently, it has been demonstrated that it is feasible to use SWE for the measurement of disc stiffness (Vergari et al. 2014), making it a very attractive tool for the exploration of pathologies and pathophysiologies related to disc degeneration and spine deformities.

10.4 APPLICATIONS IN DIFFERENT TISSUE PARTS

In addition to those tissues, which mainly include muscle, liver, breast, prostate, brain and skin as mentioned earlier, the measurement of tissue elasticity can also be applied to quite a number of other tissues including heel pad, wound, tendon, ligament, thyroid, vessel wall, gut wall, including that of oesophagus, stomach and the intestine, and the eye. Heel pad could be assessed by a portable indentation system, and the change of mechanical properties could be assessed in a special group of diabetic patients (Zheng et al. 2000a, 2012). For wounds, air-jet indentation is particularly appropriate because of its non-contact nature (Chao et al. 2011b). As tendons and ligaments are load-bearing tissues similar to muscular tissues, their mechanical properties are very important in understanding their role in locomotion and loading and related pathologies such as tendonitis (Aubry et al. 2013). Thyroid is the largest endocrine gland in the human body, and elastography is important for the differentiation between benign thyroid nodules and thyroid cancer (Andrioli and Persani 2014). The elasticity of the vessel wall is a focus of interest for the research on circulation system-related diseases, and intravascular ultrasound elastography has been developed for the assessment of vessel diseases including atherosclerosis (de Korte et al. 2000). In this scenario, the pulse wave velocity is a traditional tool for the assessment of arterial stiffness (Sutton-Tyrrell et al. 2005). However, pulse wave velocity represents the average arterial stiffness over a long region, and local measurement is still needed to characterise the local changes in arterial properties. Recent research has shown that *in vivo* measurement of arterial stiffness is possible by tracing the propagation of the acoustic radiation force-induced Lamb wave or shear wave on the vessel wall (Couade et al. 2010; Bernal et al. 2011). The digestive tract walls are of essential interest for elasticity measurement because cancer normally starts from the superficial layers. If the elasticity of the tract walls can be reliably measured, it will have the potential for the early diagnosis of digestive tissue tumours, such as gastric and colorectal cancers (Zhai et al. 2008; Chiu et al. 2009, 2012; Fakhry et al. 2009, 2014). The final application that we would like to mention here is related to the elasticity measurement of eye tissues. Because of the prevalence of ocular diseases such as myopia and keratoconus, and the involvement of change of biomechanical properties of cornea in these diseases, the measurement of corneal elasticity has received great attention in the last decade (Luce 2005). For cornea measurement, the air-puff method is especially advantageous because it is a non-contact and fast method, as implemented in the ocular response analyzer (ORA) and Scheimpflug non-contact tonometer (Luce 2005; Hon and Lam 2013). However, the parameters obtained from these two systems are currently not intrinsic mechanical properties. Once again, SWE based on acoustic radiation force stimulation and ultrasound or optical imaging has very high potential to obtain the intrinsic mechanical properties from the cornea (Manapuram et al. 2012; Nguyen et al. 2015). The trabecular meshwork and Schlemm's canal are two important components of the anterior chamber of the eyeball for controlling the outflow of the aqueous humour, which adjusts the intraocular pressure (IOP) and thus plays an important role in the pathogenesis of glaucoma. Studies have begun to focus on the investigation of the change of mechanical properties of these two tissues in regulating the IOP (Li et al. 2012b). More tissues related to the eye including the lens, retina and sclera have been the target of biomechanical studies, using bench test such as extension (Wollensak and Spoerl 2004; Hollman et al. 2007; Girard et al. 2011; Geraghty et al. 2012), although no method is available yet for the measurement of elasticity of these tissues *in vivo*.

10.5 RELATED DEVICES IN THE MARKET

In this subsection, some clinical tools existing in the market for the measurement of tissue elasticity *in vivo* are briefly introduced. Those bench testing machines that can be used for the measurement of tissue samples *in vitro* are excluded. The interested readers are strongly recommended to search the Internet for more details on the devices introduced in this subsection. These devices are divided into two types here: one that only measures the tissue elasticity as a whole, and the other that has both quantitative measurement and imaging functions.

For the indentation devices, quite a number of instruments have been commercialised as medical devices, which include the ArtScan 200 for cartilage, Cutometer® for skin and FibroScan® for the liver. ArtScan is an arthroscopic device that uses contact indentation on the articular cartilage for the measurement of its stiffness. The indentation depth is fixed, while the force gauge measures the indentation force as a reflection of the cartilage stiffness. A closely related device is the Arthro-BST™, which measures the streaming potential under compression, i.e. the mechanoelectrical properties of the articular cartilage. Myoton® was developed for the measurement of muscle properties using the resonant frequency shift method. It exerts a mechanical impulse on the skin surface over the muscles that need to be measured, and records the tissue response by an accelerometer. After computer analysis, the muscle information including tone or tension, biomechanical and viscoelastic parameters can be obtained based on information of vibration frequency change. Cutometer is a device that utilises cyclic suctions to measure the mechanical properties of the skin. Through cyclic suctions, both the elastic and viscous properties of the skin can be probed. For the corneal elasticity measurement, two devices, namely the Ocular Response Analyser (ORA) and Corvis ST, exist in the market. FibroScan is a clinical tool used for liver stiffness measurement. A shear wave is induced on the abdomen just on top of the liver, which propagates down and through the liver. The combined ultrasound transducer is used to track the propagation of the shear wave and then measures the liver stiffness, either for diagnosis or for treatment efficacy assessment. The SphygmoCor XCEL is a similar device that can measure the arterial stiffness using the pulse wave velocity measurement. The Virtual Touch Quantification is an elasticity measurement tool realised in the Siemens ACUSON ultrasound imaging system (Gallotti et al. 2010). This function is added to a conventional ultrasound imaging platform, so it has the advantage of no extra hardware purchase, lowering the cost of technology acquisition.

Based on SWE, commercial tools have been developed for the imaging of soft tissue elasticity. Aixplorer® is a supersonic shear imaging-enabled ultrasound machine for the quantitative imaging of soft tissue elasticity of various body parts. The applications of this system are very broad, because the measurement is easy and relatively user-independent, which are the main issues in the measurement of the baseline values of the elasticity of various tissues *in vivo* (Evans et al. 2010; DeWall 2013;

Gennisson et al. 2013). In this respect, MRE is another promising method that the authors are expecting to be commercially available for the measurement of important tissues such as the brain and the liver. There are also a number of strain imaging machines based on compression or acoustic radiation force impulse excitation, which provide qualitative information about soft tissue elasticity but are excluded from discussion here.

10.6 SUMMARY

Soft tissue elasticity measurement has found tremendous applications in medical and biological fields, in study of both normal tissues in physiology and simulation and pathological tissues for diagnosis and treatment assessment. From these applications, it is clear that this functional measurement of tissue elasticity has become almost a clinical tool in two particular organs, i.e. the breast and the liver, which may have significant impact on their conventional management strategies. For example, the combination of FibroScan with traditional tests significantly increases the diagnostic accuracy of liver fibrosis, and a large number of liver biopsies can be avoided as a result, although biopsy cannot be completely avoided (Sebastiani and Alberti 2006; Castera et al. 2008). For these fields with a large number of clinical trials and investigations, the future direction would be towards some guidelines that can facilitate the harmonisation of the clinical use of different systems. Some related work has already been done by the Japanese Society of Ultrasonics in Medicine (JSUM) and the European Federation of Societies for Ultrasound in Medicine and Biology (EFSUMB) (Bamber et al. 2013; Cosgrove et al. 2013; Kudo et al. 2013; Nakashima et al. 2013; Shiina 2013). For example, the classification of elastography systems, current clinical applications, important points related to elastography, practical advice, as well as the different characteristics, principles, clinical evidences and summary of various different systems are clearly presented when using the elasticity measurement techniques for breast diagnosis (Nakashima et al. 2013) in JSUM. Such types of guidelines are also given by the EFSUMB in terms of clinical applications, practical points, recommendations, pitfalls and limitations with respect to each technique used in different tissues (Cosgrove et al. 2013). For those fields where clinical instruments and data concerning measurement of mechanical properties are still inadequate, it is expected that new tools will be developed to better suit the needs of clinical requirement and push forward the related field towards a better solution. It is expected that more and more applications will be found for the measurement of soft tissue elasticity *in vivo*.

CHAPTER 11

Conclusions and Future Perspectives

11.1 SUMMARY AND TECHNICAL PERSPECTIVE

In this book, we have tried to introduce a broad range of techniques for the measurement of soft tissue elasticity *in vivo*. These techniques are mainly divided into two classes: those that achieve the task based on direct mechanical stimulation, with and without reverse engineering, and those that achieve the task through the tracing of an indirect probing medium, most likely a shear wave. Direct measurement approaches include the traditional indentation, novel indentation incorporating modern imaging methods, suction and some physiological stimulation induced by heart beats or respiration. A mechanical model of the probed tissue needs to be hypothesised, and the test can then be analysed based on continuum mechanics with appropriate boundary conditions. While theoretical analysis is possible for some simple test geometries, such as indentation on an isotropic and homogeneous thin soft tissue layer attached to an infinite, hard substrate, numerical methods such as finite element method (FEM) are a better solution with complicated tissue properties or complex testing geometries. Biological tissues are quite complex and mixed in terms of their components, showing quite obvious viscoelastic behaviour under external stimulations. The key to a successful analysis lies in a properly simplified but practical modelling of the testing behaviour using a limited number of material parameters. In this respect, the simplest material model with only two material properties, i.e. Young's modulus and Poisson's ratio, has been commonly adopted as a first step for the extraction of the material's mechanical properties. More complex models usually add a nonlinear elastic parameter or viscous parameter for a better description of the tissue behaviour. On the other hand, when the mechanical testing methods are concerned, indentation, which can be implemented for the measurement of soft tissue elasticity *in vivo*, is the simplest method. With a simple material model and indentation test, the elasticity of soft tissue can be extracted from the measured force/deformation data. Even with one parameter, the tissue elasticity measurement can be very useful in biomedical research and clinics. The elasticity value can then be used for the study of physiological changes in the tissue, such as those associated with ageing or during a certain period, or for comparison between normal and pathological states.

However, people are not satisfied with only one parameter estimation from a bulk tissue; the mechanical properties of tissues actually vary locally. What can be measured from mechanical testing methods such as indentation or suction is an overall average of material elasticity; the local variation of material properties is difficult to measure using these tests. Although it is possible to probe the local mechanical property of a tissue using a computational method such as FEM, this method is not popular because of the high computational cost and inaccurate estimation caused by too many uncertain tissue parameters. In comparison with bulk measurement, shear wave elastography (SWE), which is directly related to the tissue's shear modulus, can measure the shear wave propagation velocity in a local region. Multiple measurements of the shear modulus at different points of the tissue are thus possible using this technique, and it quickly becomes popular for clinical applications, as it is of great diagnostic value for not only diffuse diseases but also focal ones. However, the price is a high-speed ultrasound imaging system that is necessary to achieve quantitative elasticity imaging through SWE.

It is expected that future development would focus on the development of new instruments for a specific field that can achieve the measurement of tissue elasticity with less dependence on the boundary conditions and in a quick and convenient way. For SWE, the direction is towards full range, more robust, accurate and higher resolution measurement of soft tissue elasticity.

11.2 APPLICATION PERSPECTIVE

Applications associated with the measurement of soft tissue elasticity have been increasing steadily in the last two decades. Figure 11.1 shows the distribution of Science Citation Index (SCI) publication per year from

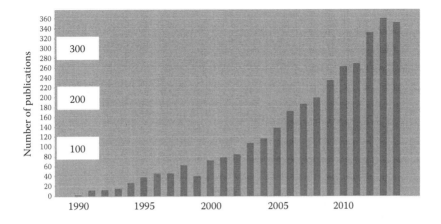

FIGURE 11.1 Distribution of publications per year (1990–2014) related to soft tissue elasticity from the Web of Science database. (From www.webofknowledge. com, accessed 21 August, 2015.)

1990 to 2014, searched by the topic 'soft tissue and (elasticity or stiffness)' in Web of Science database. (It should be noted the number of related publications is underestimated as a general term 'soft tissue' was used in the query.) It is found that the number of publications related to this area was less than 20 in the early 1990s and has increased to about 300 on average per year in the last 5 years. For those areas in which the potential of elasticity measurement has already been demonstrated by a large number of publications and clinical investigations, much more effort should be directed on how to unify the measured mechanical parameters and then incorporate the parameters as new biomarkers into conventional diagnostic process to be combined with existing markers, in order to improve the diagnostic performance and disease management strategy. Standardised training programmes or even standardised measurement procedures should be provided to doctors or clinical staff to accurately guide their operations of the related machines so that data from different measurement centres can be compared directly, thereby helping the acceleration of the technology adoption process in clinical use. On the other hand, for those areas where elasticity measurement is still at its infancy, engineers should be encouraged to collaborate more with clinical staff to conduct more clinical investigations and collect more data using those newly developed engineering tools to provide basic information in mechanical properties of the soft tissue of interest. In this way, useful feedback can be obtained from real clinical experience, which would be helpful to further improve instrument design in aspects such as ease of use, fast response, ease of disinfection and ergonomics. In this aspect, we can also expect an increase of applications related to soft tissue elasticity measurement and imaging with high resolution in small tissues such as the epidermis, epithelial gastric wall and cornea. For clinicians who want to use elasticity measurement and imaging as a clinical tool, it is very important to understand some basic physical principles of the techniques as well as the assumptions and constraints of each technique or device. We also hope that the terminology used in the field will become more standardised as time goes, and more intrinsic material parameters will be used.

References

Addae-Mensah, K. A. and J. P. Wikswo. 2008. Measurement techniques for cellular biomechanics in vitro. *Experimental Biology and Medicine* 233 (7):792–809.

Aglyamov, S. R., A. B. Karpiouk, M. Mehrmohammadi, S. Yoon, S. Kim, Y. A. Ilinskii, E. A. Zabolotskaya and S. Y. Emelianov. 2012. Elasticity imaging and sensing using targeted motion: From macro to nano. *Current Medical Imaging Reviews* 8 (1):3–15.

Alblas, J. B. and M. Kuipers. 1970. Contact problems of a rectangular block on an elastic layer of finite thickness. Part II: The thick layer. *Acta Mechanica* 9 (1–2):1–12.

Alexander, H. and T. H. Cook. 1977. Accounting for natural tension in the mechanical testing of human skin. *Journal of Investigative Dermatology* 69 (3):310–314.

Alexopoulos, L. G., M. A. Haider, T. P. Vail and F. Guilak. 2003. Alterations in the mechanical properties of the human chondrocyte pericellular matrix with osteoarthritis. *Journal of Biomechanical Engineering – Transactions of the ASME* 125 (3):323–333.

Almeida-Silveira, M. I., D. Lambertz, C. Perot and F. Goubel. 2000. Changes in stiffness induced by hindlimb suspension in rat Achilles tendon. *European Journal of Applied Physiology* 81 (3):252–257.

Amador, C., M. W. Urban, S. G. Chen, Q. S. Chen, K. N. An and J. F. Greenleaf. 2011. Shear elastic modulus estimation from indentation and SDUV on gelatin phantoms. *IEEE Transactions on Biomedical Engineering* 58 (6):1706–1714.

Andreassen, T. T., A. H. Simonsen and H. Oxlund. 1980. Biomechanical properties of keratoconus and normal corneas. *Experimental Eye Research* 31 (4):435–441.

Andrioli, M. and L. Persani. 2014. Elastographic techniques of thyroid gland: Current status. *Endocrine* 46 (3):455–461.

Angker, L. and M. V. Swain. 2006. Nanoindentation: Application to dental hard tissue investigations. *Journal of Materials Research* 21 (8):1893–1905.

Aoki, T., T. Ohashi, T. Matsumoto and M. Sato. 1997. The pipette aspiration applied to the local stiffness measurement of soft tissues. *Annals of Biomedical Engineering* 25 (3):581–587.

Arda, K., N. Ciledag, E. Aktas, B. K. Aribas and K. Kose. 2011. Quantitative assessment of normal soft-tissue elasticity using shear-wave ultrasound elastography. *American Journal of Roentgenology* 197 (3):532–536.

Armstrong, C. G., A. S. Bahrani and D. L. Gardner. 1979. In vitro measurement of articular cartilage deformations in the intact human hip joint under load. *Journal of Bone and Joint Surgery – American Volume* 61 (5):744–755.

Ateshian, G. A., W. H. Warden, J. J. Kim, R. P. Grelsamer and V. C. Mow. 1997. Finite deformation biphasic material properties of bovine articular cartilage from confined compression experiments. *Journal of Biomechanics* 30 (11–12):1157–1164.

Athanasiou, A., A. Tardivon, M. Tanter, B. Sigal-Zafrani, J. Bercoff, T. D. Eux, J. L. Gennisson, M. Fink and S. Neuenschwander. 2010. Breast lesions: Quantitative elastography with supersonic shear imaging-preliminary results. *Radiology* 256 (1):297–303.

Athanasiou, K. A., A. Agarwal and F. J. Dzida. 1994. Comparative study of the intrinsic mechanical properties of the human acetabular and femoral head cartilage. *Journal of Orthopaedic Research* 12 (3):340–349.

Athanasiou, K. A., A. Agarwal, A. Muffoletto, F. J. Dzida, G. Constantinides and M. Clem. 1995. Biomechanical properties of hip cartilage in experimental animal models. *Clinical Orthopaedics and Related Research* 316:254–266.

Athanasiou, K. A., M. P. Rosenwasser, J. A. Buckwalter, T. I. Malinin and V. C. Mow. 1991. Interspecies comparisons of in situ intrinsic mechanical properties of distal femoral cartilage. *Journal of Orthopaedic Research* 9 (3):330–340.

Aubry, S., J. R. Risson, A. Kastler, B. Barbier-Brion, G. Siliman, M. Runge and B. Kastler. 2013. Biomechanical properties of the calcaneal tendon in vivo assessed by transient shear wave elastography. *Skeletal Radiology* 42 (8):1143–1150.

Avenhaus, W., B. Kemper, G. von Bally and W. Domschke. 2001. Gastric wall elasticity assessed by dynamic holographic endoscopy: *Ex vivo* investigations in the porcine stomach. *Gastrointestinal Endoscopy* 54 (4):496–500.

Avikainen, V. J., A. Rezasoltani and H. A. Kauhanen. 1999. Asymmetry of paraspinal EMG-time characteristics in idiopathic scoliosis. *Journal of Spinal Disorders* 12 (1):61–67.

Bader, D. L. and P. Bowker. 1983. Mechanical characteristics of skin and underlying tissues in vivo. *Biomaterials* 4 (4):305–308.

Bae, U., M. Dighe, T. Dubinsky, S. Minoshima, V. Shamdasani and Y. M. Kim. 2007. Ultrasound thyroid elastography using carotid artery pulsation – Preliminary study. *Journal of Ultrasound in Medicine* 26 (6):797–805.

Balleyguier, C., S. Canale, W. Ben Hassen, P. Vielh, E. H. Bayou, M. C. Mathieu, C. Uzan, C. Bourgier and C. Dromain. 2013. Breast elasticity: Principles, technique, results: An update and overview of commercially available software. *European Journal of Radiology* 82 (3):427–434.

Bamber, J., D. Cosgrove, C. F. Dietrich, J. Fromageau, J. Bojunga, F. Calliada, V. Cantisani et al. 2013. EFSUMB guidelines and recommendations on the clinical use of ultrasound elastography. Part 1: Basic principles and technology. *Ultraschall in Der Medizin* 34 (2):169–184.

Bank, R. A., M. Soudry, A. Maroudas, J. Mizrahi and J. M. TeKoppele. 2000. The increased swelling and instantaneous deformation of osteoarthritic cartilage is highly correlated with collagen degradation. *Arthritis and Rheumatism* 43 (10):2202–2210.

Barbarino, G. G., M. Jabareen and E. Mazza. 2011. Experimental and numerical study on the mechanical behavior of the superficial layers of the face. *Skin Research and Technology* 17 (4):434–444.

Barbone, P. E. and A. A. Oberai. 2010. A review of the mathematical and computational foundations of biomechanical imaging. In *Computational Modeling in Biomechanics*, eds. S. De, F. Guilak and M. R. K. Mofrad. Berlin, Germany: Springer-Verlag.

Basser, P. J., R. Schneiderman, R. A. Bank, E. Wachtel and A. Maroudas. 1998. Mechanical properties of the collagen network in human articular cartilage as measured by osmotic stress technique. *Archives of Biochemistry and Biophysics* 351 (2):207–219.

Bauer, M., E. Mazza, M. Jabareen, L. Sultan, M. Bajka, U. Lang, R. Zimmermann and G. A. Holzapfel. 2009. Assessment of the in vivo biomechanical properties of the human uterine cervix in pregnancy using the aspiration test – A feasibility study. *European Journal of Obstetrics & Gynecology and Reproductive Biology* 144:S77–S81.

Beer, F. P., E. R. Johnston and J. T. Dewolf. 2006a. Torsion. In *Mechanics of Materials*, eds. F. P. Beer, E. R. Johnston and J. T. Dewolf. Singapore: McGraw Hill, Chapter 2.

Beer, F. P., E. R. Johnston and J. T Dewolf. 2006b. Pure bending. In *Mechanics of Materials*, eds. F. P. Beer, E. R. Johnston and J. T. Dewolf. Singapore: McGraw Hill, Chapter 3.

Bercoff, J., M. Tanter and M. Fink. 2004a. Supersonic shear imaging: A new technique for soft tissue elasticity mapping. *IEEE Transactions on Ultrasonics Ferroelectrics and Frequency Control* 51 (4):396–409.

Bercoff, J., M. Tanter, M. Muller and M. Fink. 2004b. The role of viscosity in the impulse diffraction field of elastic waves induced by the acoustic radiation force. *IEEE Transactions on Ultrasonics Ferroelectrics and Frequency Control* 51 (11):1523–1536.

Bernal, M., I. Nenadic, M. W. Urban and J. F. Greenleaf. 2011. Material property estimation for tubes and arteries using ultrasound radiation force and analysis of propagating modes. *Journal of the Acoustical Society of America* 129 (3):1344–1354.

Bosetti, C., F. Levi, F. Lucchini, W. A. Zatonski, E. Negri and C. La Vecchia. 2007. Worldwide mortality from cirrhosis: An update to 2002. *Journal of Hepatology* 46 (5):827–839.

Bouchard, R. R., J. J. Dahl, S. J. Hsu, M. L. Palmeri and G. E. Trahey. 2009a. Image quality, tissue heating, and frame rate trade-offs in acoustic radiation force impulse imaging. *IEEE Transactions on Ultrasonics Ferroelectrics and Frequency Control* 56 (1):63–76.

Bouchard, R. R., M. L. Palmeri, G. F. Pinton, G. E. Trahey, J. E. Streeter and P. A. Dayton. 2009b. Optical tracking of acoustic radiation force impulse-induced dynamics in a tissue-mimicking phantom. *Journal of the Acoustical Society of America* 126 (5):2733–2745.

Boudou, T., J. Ohayon, Y. Arntz, G. Finet, C. Picart and P. Tracqui. 2006. An extended modeling of the micropipette aspiration experiment for the characterization of the Young's modulus and Poisson's ratio of adherent thin biological samples: Numerical and experimental studies. *Journal of Biomechanics* 39 (9):1677–1685.

Bouillard, K., A. Nordez and F. Hug. 2011. Estimation of individual muscle force using elastography. *PLoS ONE* 6 (12):e29261.

Brandenburg, J. E., S. F. Eby, P. F. Song, H. Zhao, J. S. Brault, S. G. Chen and K. N. An. 2014. Ultrasound elastography: The new frontier in direct measurement of muscle stiffness. *Archives of Physical Medicine and Rehabilitation* 95 (11):2207–2219.

Briscoe, B. J., L. Fiori and E. Pelillo. 1998. Nano-indentation of polymeric surfaces. *Journal of Physics D – Applied Physics* 31 (19):2395–2405.

Burd, H. J., G. S. Wilde and S. J. Judge. 2011. An improved spinning lens test to determine the stiffness of the human lens. *Experimental Eye Research* 92 (1):28–39.

Carrillo, F., S. Gupta, M. Balooch, S. J. Marshall, G. W. Marshall, L. Pruitt and C. M. Puttlitz. 2005. Nanoindentation of polydimethylsiloxane elastomers: Effect of crosslinking, work of adhesion, and fluid environment on elastic modulus. *Journal of Materials Research* 20 (10):2820–2830.

Castaneda, B., J. Ormachea, P. Rodriguez and K. J. Parker. 2013. Application of numerical methods to elasticity imaging. *Molecular & Cellular Biomechanics* 10 (1):43–65.

Castera, L. 2011. Non-invasive assessment of liver fibrosis in chronic hepatitis C. *Hepatology International* 5 (2):625–634.

Castera, L., X. Forns and A. Alberti. 2008. Non-invasive evaluation of liver fibrosis using transient elastography. *Journal of Hepatology* 48 (5):835–847.

Castera, L., J. Vergniol, J. Foucher, B. Le Bail, E. Chanteloup, M. Haaser, M. Darriet, P. Couzigou and V. De Ledinghen. 2005. Prospective comparison of transient elastography, fibrotest, APRI, and liver biopsy for the assessment of fibrosis in chronic hepatitis C. *Gastroenterology* 128 (2):343–350.

Catheline, S., J. L. Gennisson, G. Delon, M. Fink, R. Sinkus, S. Abouelkaram and J. Culioli. 2004. Measurement of viscoelastic properties of homogeneous soft solid using transient elastography: An inverse problem approach. *Journal of the Acoustical Society of America* 116 (6):3734–3741.

Catheline, S., J. L. Thomas, F. Wu and M. A. Fink. 1999a. Diffraction field of a low frequency vibrator in soft tissues using transient elastography. *IEEE Transactions on Ultrasonics Ferroelectrics and Frequency Control* 46 (4):1013–1019.

Catheline, S., F. Wu and M. Fink. 1999b. A solution to diffraction biases in sono-elasticity: The acoustic impulse technique. *Journal of the Acoustical Society of America* 105 (5):2941–2950.

Cespedes, I., J. Ophir, H. Ponnekanti and N. Maklad. 1993. Elastography – Elasticity imaging using ultrasound with application to muscle and breast in vivo. *Ultrasonic Imaging* 15 (2):73–88.

Chai, C. K., H. J. Burd and G. S. Wilde. 2012. Shear modulus measurements on isolated human lens nuclei. *Experimental Eye Research* 103:78–81.

Chang, J. M., W. K. Moon, N. Cho, A. Yi, H. R. Koo, W. Han, D. Y. Noh, H. G. Moon and S. J. Kim. 2011. Clinical application of shear wave elastography (SWE) in the diagnosis of benign and malignant breast diseases. *Breast Cancer Research and Treatment* 129 (1):89–97.

Chao, C. Y. L., G. Y. F. Ng, K. K. Cheung, Y. P. Zheng, L. K. Wang and G. L. Y. Cheing. 2013. In vivo and ex vivo approaches to studying the biomechanical properties of healing wounds in rat skin. *Journal of Biomechanical Engineering – Transactions of the ASME* 135 (10):101009.

Chao, C. Y. L., Y. P. Zheng and G. L. Y. Cheing. 2011a. Epidermal thickness and biomechanical properties of plantar tissue in diabetic foot. *Ultrasound in Medicine and Biology* 37 (7):1029–1038.

Chao, C. Y. L., Y. P. Zheng, Y. P. Huang and G. L. Y. Cheing. 2010. Biomechanical properties of the forefoot plantar soft tissue as measured by an optical coherence tomography-based air-jet indentation system and tissue ultrasound palpation system. *Clinical Biomechanics* 25 (6):594–600.

Chao, C. Y. L., Y. P. Zheng and G. L. Y. Cheing. 2011b. A novel noncontact method to assess the biomechanical properties of wound tissue. *Wound Repair and Regeneration* 19 (3):324–329.

Chen, H. C., B. M. O'Brien, J. J. Pribaz and A. H. N. Roberts. 1988. The use of tonometry in the assessment of upper extremity lymphoedema. *British Journal of Plastic Surgery* 41 (4):399–402.

Chen, S. G., M. Fatemi and J. F. Greenleaf. 2004. Quantifying elasticity and viscosity from measurement of shear wave speed dispersion. *Journal of the Acoustical Society of America* 115 (6):2781–2785.

Chen, S. G., M. W. Urban, C. Pislaru, R. Kinnick, Y. Zheng, A. P. Yao and J. F. Greenleaf. 2009. Shearwave dispersion ultrasound vibrometry (SDUV) for measuring tissue elasticity and viscosity. *IEEE Transactions on Ultrasonics Ferroelectrics and Frequency Control* 56 (1):55–62.

Chen, S. S., Y. H. Falcovitz, R. Schneiderman, A. Maroudas and R. L. Sah. 2001. Depth-dependent compressive properties of normal aged human femoral head articular cartilage: Relationship to fixed charge density. *Osteoarthritis and Cartilage* 9 (6):561–569.

Cheng, Y. T., W. Y. Ni and C. M. Cheng. 2005. Determining the instantaneous modulus of viscoelastic solids using instrumented indentation measurements. *Journal of Materials Research* 20 (11):3061–3071.

Chieffi, M. 1950. An investigation of the effects of parenteral and topical administration of steroids on the elastic properties of senile skin. *Journal of Gerontology* 5 (1):17–22.

Chien, S., K. L. P. Sung, R. Skalak and S. Usami. 1978. Theoretical and experimental studies on viscoelastic properties of erythrocyte membrane. *Biophysical Journal* 24 (2):463–487.

Chiu, P. W., Y. P. Zheng, Y. P. Huang, K. N. Enders, K. F. To, S. K. Kong and A. H. Ho. May 30–Jun 04 2009. Development of endoluminal technique for detection of elasticity in gastrointestinal tract during endoscopy. Paper read at Digestive Disease Week Meeting/110th Annual Meeting of the American Society for Gastrointestinal Endoscopy, Chicago, IL.

Chiu, P. W., Y. P. Zheng, Y. P. Huang, L. K. Wang, A. H. Ho, S. Kong, K. F. To and F. K. L. Chan. 2012. Development of a novel endoluminal pressure system for detection of changes in elasticity for recognition of gastrointestinal neoplasia. *Gastrointestinal Endoscopy* 75 (4):482–482.

Choi, A. P. C. 2009. Estimation of Young's modulus and Poisson's ratio of soft tissue Using indentation. MPhil, Health Technology and Informatics, Hong Kong Polytechnic University, Hung Hom, Hong Kong.

Choi, A. P. C. and Y. P. Zheng. 2005. Estimation of Young's modulus and Poisson's ratio of soft tissue from indentation using two different-sized indentors: Finite element analysis of the finite deformation effect. *Medical & Biological Engineering & Computing* 43 (2):258–264.

Cobbold, J. F. L. and S. D. Taylor-Robinson. 2008. Transient elastography in acute hepatitis: All that's stiff is not fibrosis. *Hepatology* 47 (2):370–372.

Cook, C. S. and M. J. N. McDonagh. 1996. Measurement of muscle and tendon stiffness in man. *European Journal of Applied Physiology and Occupational Physiology* 72 (4):380–382.

Cook, R. F. 2010. Probing the nanoscale. *Science* 328 (5975):183–184.

Cook, T., H. Alexander and M. Cohen. 1977. Experimental method for determining 2-dimensional mechanical properties of living human skin. *Medical & Biological Engineering & Computing* 15 (4):381–390.

Corica, G. F., N. C. Wigger, D. W. Edgar, F. M. Wood and S. Carroll. 2006. Objective measurement of scarring by multiple assessors: Is the tissue tonometer a reliable option? *Journal of Burn Care & Research* 27 (4):520–523.

Corpechot, C., F. Carrat, A. Poujol-Robert, F. Gaouar, D. Wendum, O. Chazouilleres and R. Poupon. 2012. Noninvasive elastography-based assessment of liver fibrosis progression and prognosis in primary biliary cirrhosis. *Hepatology* 56 (1):198–208.

Cortes, D. H., J. F. Magland, A. C. Wright and D. M. Elliott. 2014. The shear modulus of the nucleus pulposus measured using magnetic resonance elastography: A potential biomarker for intervertebral disc degeneration. *Magnetic Resonance in Medicine* 72 (1):211–219.

Cosgrove, D., F. Piscaglia, J. Bamber, J. Bojunga, J. M. Correas, O. H. Gilja, A. S. Klauser et al. 2013. EFSUMB guidelines and recommendations on the clinical use of ultrasound elastography. Part 2: Clinical applications. *Ultraschall in Der Medizin* 34 (3):238–253.

Couade, M., M. Pernot, C. Prada, E. Messas, J. Emmerich, A. Bruneval, A. Criton, M. Fink and M. Tanter. 2010. Quantitative assessment of arterial wall biomechanical properties using shear wave imaging. *Ultrasound in Medicine and Biology* 36 (10):1662–1676.

Cuy, J. L., A. B. Mann, K. J. Livi, M. F. Teaford and T. P. Weihs. 2002. Nanoindentation mapping of the mechanical properties of human molar tooth enamel. *Archives of Oral Biology* 47 (4):281–291.

de Korte, C. L., S. G. Carlier, F. Mastik, M. M. Doyley, A. F. W. van der Steen, P. W. Serruys and N. Bom. 2002. Morphological and mechanical information of coronary arteries obtained with intravascular elastography – Feasibility study in vivo. *European Heart Journal* 23 (5):405–413.

de Korte, C. L., E. I. Cespedes, A. F. W. vanderSteen and C. T. Lancee. 1997. Intravascular elasticity imaging using ultrasound: Feasibility studies in phantoms. *Ultrasound in Medicine and Biology* 23 (5):735–746.

de Korte, C. L., G. Pasterkamp, A. F. W. van der Steen, H. A. Woutman and N. Bom. 2000. Characterization of plaque components with intravascular ultrasound elastography in human femoral and coronary arteries in vitro. *Circulation* 102 (6):617–623.

de Korte, C. L. and A. F. W. van der Steen. 2002. Intravascular ultrasound elastography: An overview. *Ultrasonics* 40:859–865.

Deffieux, T., G. Montaldo, M. Tanter and M. Fink. 2009. Shear wave spectroscopy for in vivo quantification of human soft tissues visco-elasticity. *IEEE Transactions on Medical Imaging* 28 (3):313–322.

DeWall, R. J. 2013. Ultrasound elastography: Principles, techniques, and clinical applications. *Critical Reviews in Biomedical Engineering* 41 (1):1–19.

Dhanaliwala, A. H., J. A. Hossack and F. W. Mauldin. 2012. Assessing and improving acoustic radiation force image quality using a 1.5-D transducer design. *IEEE Transactions on Ultrasonics Ferroelectrics and Frequency Control* 59 (7):1602–1608.

Di Ieva, A., F. Grizzi, E. Rognone, Z. T. H. Tse, T. Parittotokkaporn, F. R. Y. Baena, M. Tschabitscher, C. Matula, S. Trattnig and R. R. Y. Baena. 2010. Magnetic resonance elastography: A general overview of its current and future applications in brain imaging. *Neurosurgical Review* 33 (2):137–145.

Diridollou, S., F. Patat, F. Gens, L. Vaillant, D. Black, J. M. Lagarde, Y. Gall and M. Berson. 2000. In vivo model of the mechanical properties of the human skin under suction. *Skin Research and Technology* 6 (4):214–221.

Discher, D. E., N. Mohandas and E. A. Evans. 1994. Molecular maps of red-cell deformation – Hidden elasticity and in-situ connectivity. *Science* 266 (5187):1032–1035.

Dobrev, H. 2005. Application of Cutometer area parameters for the study of human skin fatigue. *Skin Research and Technology* 11 (2):120–122.

Dokos, S., I. J. LeGrice, B. H. Smaill, J. Kar and A. A. Young. 2000. A triaxial-measurement shear-test device for soft biological tissues. *Journal of Biomechanical Engineering – Transactions of the ASME* 122 (5):471–478.

Dokos, S., B. H. Smaill, A. A. Young and I. J. LeGrice. 2002. Shear properties of passive ventricular myocardium. *American Journal of Physiology – Heart and Circulatory Physiology* 283 (6):H2650–H2659.

Doyley, M. M. 2012. Model-based elastography: A survey of approaches to the inverse elasticity problem. *Physics in Medicine and Biology* 57 (3):R35–R73.

Drager, L. F., L. Diegues-Silva, P. M. Diniz, G. Lorenzi, E. M. Krieger and L. A. Bortolotto. 2011. Lack of circadian variation of pulse vave velocity measurements in healthy volunteers. *Journal of Clinical Hypertension* 13 (1):19–22.

Drapaca, C. S., S. Sivaloganathan and G. Tenti. 2007. Nonlinear constitutive laws in viscoelasticity. *Mathematics and Mechanics of Solids* 12 (5):475–501.

Dubbelman, M., H. A. Weeber, R. G. L. van der Heijde and H. J. Volker-Dieben. 2002. Radius and asphericity of the posterior corneal surface determined by corrected Scheimpflug photography. *Acta Ophthalmologica Scandinavica* 80 (4):379–383.

Duda, G. N., R. U. Kleemann, U. Bluecher and A. Weiler. 2004. A new device to detect early cartilage degeneration. *American Journal of Sports Medicine* 32 (3):693–698.

Dugar, M., R. Woolford, M. J. Ahern, M. D. Smith and P. J. Roberts-Thomson. 2009. Use of electronic tonometer to assess skin hardness in systemic sclerosis: A pilot cross-sectional study. *Clinical and Experimental Rheumatology* 27 (3):S70–S70.

Ebenstein, D. M., D. Coughlin, J. Chapman, C. Li and L. A. Pruitt. 2009. Nanomechanical properties of calcification, fibrous tissue, and hematoma from atherosclerotic plaques. *Journal of Biomedical Materials Research Part A* 91A (4):1028–1037.

Ebenstein, D. M., A. Kuo, J. J. Rodrigo, A. H. Reddi, M. Ries and L. Pruitt. 2004. A nanoindentation technique for functional evaluation of cartilage repair tissue. *Journal of Materials Research* 19 (1):273–281.

Ebenstein, D. M. and L. A. Pruitt. 2006. Nanoindentation of biological materials. *Nano Today* 1 (3):26–33.

Ebenstein, D. M. and K. J. Wahl. 2006. A comparison of JKR-based methods to analyze quasi-static and dynamic indentation force curves. *Journal of Colloid and Interface Science* 298 (2):652–662.

Eder, A., M. Richter, C. Kargel and IEEE. 2012. A new approach to improve the reconstruction quality in ultrasound elastography. In *2012 IEEE International Instrumentation and Measurement Technology Conference (I2 MTC)*, Graz, Austria, pp. 1908–1913.

Eklund, A., A. Bergh and O. A. Lindahl. 1999. A catheter tactile sensor for measuring hardness of soft tissue: Measurement in a silicone model and in an in vitro human prostate model. *Medical & Biological Engineering & Computing* 37 (5):618–624.

Eklund, A., P. Hallberg, C. Linden and O. A. Lindahl. 2003. An applanation resonator sensor for measuring intraocular pressure using combined continuous force and area measurement. *Investigative Ophthalmology & Visual Science* 44 (7):3017–3024.

Elsheikh, A., D. F. Wang, M. Brown, P. Rama, M. Campanelli and D. Pye. 2007. Assessment of corneal biomechanical properties and their variation with age. *Current Eye Research* 32 (1):11–19.

Elsner, P., D. Wilhelm and H. I. Maibach. 1990. Mechanical properties of human forearm and vulvar skin. *British Journal of Dermatology* 122 (5): 607–614.

Engler, A. J., S. Sen, H. L. Sweeney and D. E. Discher. 2006. Matrix elasticity directs stem cell lineage specification. *Cell* 126 (4):677–689.

Enomoto, D. N. H., J. R. Mekkes, P. M. M. Bossuyt, R. Hoekzema and J. D. Bos. 1996. Quantification of cutaneous sclerosis with a skin elasticity meter in patients with generalized scleroderma. *Journal of the American Academy of Dermatology* 35 (3):381–387.

Eshel, H. and Y. Lanir. 2001. Effects of strain level and proteoglycan depletion on preconditioning and viscoelastic responses of rat dorsal skin. *Annals of Biomedical Engineering* 29 (2):164–172.

Evans, A., P. Whelehan, K. Thomson, D. McLean, K. Brauer, C. Purdie, L. Jordan, L. Baker and A. Thompson. 2010. Quantitative shear wave ultrasound elastography: Initial experience in solid breast masses. *Breast Cancer Research* 12 (6):R104.

Evans, E. and A. Yeung. 1989. Apparent viscosity and cortical tension of blood granulocytes determined by micropipet aspiration. *Biophysical Journal* 56 (1):151–160.

Fahey, B. J., K. R. Nightingale, R. C. Nelson, M. L. Palmeri and G. E. Trahey. 2005. Acoustic radiation force impulse imaging of the abdomen: Demonstration of feasibility and utility. *Ultrasound in Medicine and Biology* 31 (9):1185–1198.

Fakhry, M., F. Bello and G. B. Hanna. 2009. A real-time compliance mapping system using standard endoscopic surgical forceps. *IEEE Transactions on Biomedical Engineering* 56 (4):1245–1253.

Fakhry, M., F. Bello and G. B. Hanna. 2014. Real time cancer prediction based on objective tissue compliance measurement in endoscopic surgery. *Annals of Surgery* 259 (2):369–373.

Falanga, V. and B. Bucalo. 1993. Use of a durometer to assess skin hardness. *Journal of the American Academy of Dermatology* 29 (1):47–51.

Fatemi, M. and J. F. Greenleaf. 1998. Ultrasound-stimulated vibro-acoustic spectrography. *Science* 280 (5360):82–85.

Fatemi, M. and J. F. Greenleaf. 1999a. Application of radiation force in noncontact measurement of the elastic parameters. *Ultrasonic Imaging* 21 (2):147–154.

Fatemi, M. and J. F. Greenleaf. 1999b. Vibro-acoustography: An imaging modality based on ultrasound-stimulated acoustic emission. *Proceedings of the National Academy of Sciences of the United States of America* 96 (12):6603–6608.

Fatemi, M. and J. F. Greenleaf. 2002. Imaging the viscoelastic properties of tissue. In *Imaging of Complex Media with Acoustic and Seismic Waves*. Berlin, Germany: Springer-Verlag.

Fatemi, M., L. E. Wold, A. Alizad and J. F. Greenleaf. 2002. Vibro-acoustic tissue mammography. *IEEE Transactions on Medical Imaging* 21 (1):1–8.

Feng, G. and A. H. W. Ngan. 2002. Effects of creep and thermal drift on modulus measurement using depth-sensing indentation. *Journal of Materials Research* 17 (3):660–668.

Ferguson-Pell, M., S. Hagisawa and R. D. Masiello. 1994. A skin indentation system using a pneumatic bellows. *Journal of Rehabilitation Research and Development* 31 (1):15–19.

Ferguson, V. L., A. J. Bushby and A. Boyde. 2003. Nanomechanical properties and mineral concentration in articular calcified cartilage and subchondral bone. *Journal of Anatomy* 203 (2):191–202.

Fink, M. and M. Tanter. 2011. A multiwave imaging approach for elastography. *Current Medical Imaging Reviews* 7 (4):340–349.

Fischer, A. A. 1987a. Clinical use of tissue compliance meter for documentation of soft tissue pathology. *Clinical Journal of Pain* 3 (1):23–30.

Fischer, A. A. 1987b. Tissue compliance meter for objective, quantitative documentation of soft-tissue consistency and pathology. *Archives of Physical Medicine and Rehabilitation* 68 (2):122–125.

Fischer-Cripps, A. C. 2006. Review of analysis and interpretation of nanoindentation test data. *Surface & Coatings Technology* 200 (14–15):4153–4165.

Fisher, R. F. 1971. The elastic constants of the human lens. *Journal of Physiology* 212 (1):147–180.

Flahiff, C. M., V. B. Kraus, J. L. Huebner and L. A. Setton. 2004. Cartilage mechanics in the guinea pig model of osteoarthritis studied with an osmotic loading method. *Osteoarthritis and Cartilage* 12 (5):383–388.

Flahiff, C. M., D. A. Narmoneva, J. L. Huebner, V. B. Kraus, F. Guilak and L. A. Setton. 2002. Osmotic loading to determine the intrinsic material properties of guinea pig knee cartilage. *Journal of Biomechanics* 35 (9):1285–1290.

Fontes, B. M., R. Ambrosio, G. C. Velarde and W. Nose. 2011. Ocular response analyzer measurements in keratoconus with normal central corneal thickness compared with matched normal control eyes. *Journal of Refractive Surgery* 27 (3):209–215.

Fortin, M., M. D. Buschmann, M. J. Bertrand, F. S. Foster, and J. Ophir. 2003. Dynamic measurement of internal solid displacement in articular cartilage using ultrasound backscatter. *Journal of Biomechanics* 36 (3):443–447.

Foucher, J., E. Chanteloup, J. Vergniol, L. Castera, B. Le Bail, X. Adhoute, J. Bertet, P. Couzigou and V. de Ledinghen. 2006. Diagnosis of cirrhosis by transient elastography (FibroScan): A prospective study. *Gut* 55 (3):403–408.

Franke, O., K. Durst, V. Maier, M. Goken, T. Birkhoiz, H. Schneider, F. Hennig and K. Gelse. 2007. Mechanical properties of hyaline and repair cartilage studied by nanoindentation. *Acta Biomaterialia* 3 (6):873–881.

Franke, O., M. Goken and A. M. Hodge. 2008. The nanoindentation of soft tissue: Current and developing approaches. *JOM* 60 (6):49–53.

Franke, O., M. Goken, M. A. Meyers, K. Durst and A. M. Hodge. 2011. Dynamic nanoindentation of articular porcine cartilage. *Materials Science & Engineering C-Materials for Biological Applications* 31 (4):789–795.

Friedman, S. L. 2003. Liver fibrosis – From bench to bedside. *Journal of Hepatology* 38:S38–S53.

Friedrich-Rust, M., M. F. Ong, S. Martens, C. Sarrazin, J. Bojunga, S. Zeuzem and E. Herrmann. 2008. Performance of transient elastography for the staging of liver fibrosis: A meta-analysis. *Gastroenterology* 134 (4):960–974.

Fung, Y. C. 1965. Elastic and plastic behavior of materials. In *Foundations of Solid Mechanics*, ed. Y. C. Fung. Englewood Cliffs, NJ: Prentice Hall, Chapter 6.

Fung, Y. C. 1993a. *Biomechanics: Mechanical Properties of Living Tissues*. New York: Springer.

Fung, Y. C. 1993b. The meaning of the constitutive equation. In *Biomechanics: Mechanical Properties of Living Tissues*, ed. Y. C. Fung. New York: Springer-Verlag, Chapter 2.

Fung, Y. C. 1993c. Bioviscoelastic solids. In *Biomechanics: Mechanical Properties of Living Tissues*, ed. Y. C. Fung. New York: Springer-Verlag, Chapter 7.

Fung, Y. C. 1993d. *First Course in Continuum Mechanics*, 3rd edn. Englewood Cliffs, NJ: Prentice-Hall Inc.

Gallotti, A., M. D'Onofrio and R. P. Mucelli. 2010. Acoustic radiation force impulse (ARFI) technique in ultrasound with virtual touch tissue quantification of the upper abdomen. *Radiologia Medica* 115 (6):889–897.

Gao, L., K. J. Parker, S. K. Alam and R. M. Lerner. 1995. Sonoelasticity imaging – Theory and experimental verification. *Journal of the Acoustical Society of America* 97 (6):3875–3886.

Gao, L., K. J. Parker, R. M. Lerner and S. F. Levinson. 1996. Imaging of the elastic properties of tissue – A review. *Ultrasound in Medicine and Biology* 22 (8):959–977.

Gasser, T. C., R. W. Ogden and G. A. Holzapfel. 2006. Hyperelastic modelling of arterial layers with distributed collagen fibre orientations. *Journal of the Royal Society Interface* 3 (6):15–35.

Gefen, A., M. Megido-Ravid, M. Azariah, Y. Itzchak and M. Arcan. 2001a. Integration of plantar soft tissue stiffness measurements in routine MRI of the diabetic foot. *Clinical Biomechanics* 16 (10):921–925.

Gefen, A., M. Megido-Ravid and Y. Itzchak. 2001b. In vivo biomechanical behavior of the human heel pad during the stance phase of gait. *Journal of Biomechanics* 34 (12):1661–1665.

Gennisson, J. L., T. Deffieux, M. Fink and M. Tanter. 2013. Ultrasound elastography: Principles and techniques. *Diagnostic and Interventional Imaging* 94 (5):487–495.

Gennisson, J. L., T. Deffieux, E. Mace, G. Montaldo, M. Fink and M. Tanter. 2010. Viscoelastic and anisotropic mechanical properties of *in vivo* muscle tissue assessed by supersonic shear imaging. *Ultrasound in Medicine and Biology* 36 (5):789–801.

Gentleman, E., A. N. Lay, D. A. Dickerson, E. A. Nauman, G. A. Livesay and K. C. Dee. 2003. Mechanical characterization of collagen fibers and scaffolds for tissue engineering. *Biomaterials* 24 (21):3805–3813.

Geraghty, B., S. W. Jones, P. Rama, R. Akhtar and A. Elsheikh. 2012. Age-related variations in the biomechanical properties of human sclera. *Journal of the Mechanical Behavior of Biomedical Materials* 16:181–191.

Girard, M. J. A., J. K. F. Suh, M. Bottlang, C. F. Burgoyne and J. C. Downs. 2011. Biomechanical changes in the sclera of monkey eyes exposed to chronic IOP elevations. *Investigative Ophthalmology & Visual Science* 52 (8):5656–5669.

Gniadecka, M. and J. Serup. 2006. Suction chamber method for measuring skin mechanical properties: The Dermaflex®. In *Handbook of Non-Invasive Methods and the Skin*, eds. J. Serup and G. Jemec. Boca Raton, FL: CRC, Chapter 65.

Goenezen, S., J. F. Dord, Z. Sink, P. E. Barbone, J. Jiang, T. J. Hall and A. A. Oberai. 2012. Linear and nonlinear elastic modulus imaging: An application to breast cancer diagnosis. *IEEE Transactions on Medical Imaging* 31 (8):1628–1637.

Goldich, Y., Y. Barkana, Y. Morad, M. Hartstein, I. Avni and D. Zadok. 2009. Can we measure corneal biomechanical changes after collagen cross-linking in eyes with keratoconus? A pilot study. *Cornea* 28 (5):498–502.

Goldmann, H. and T. Schmidt. 1957. Applanation tonometry. *Ophthalmologica* 134 (4):221–242.

Good, D. W., G. D. Stewart, S. Hammer, P. Scanlan, W. Shu, S. Phipps, R. Reuben and A. S. McNeill. 2014. Elasticity as a biomarker for prostate cancer: A systematic review. *BJU International* 113 (4):523–534.

Grahame, R. 1970. A method for measuring human skin elasticity in vivo with observations of the effects of age, sex and pregnancy. *Clinical Science* 39 (2):223–229.

Grahame, R. and P. J. L. Holt. 1969. The influence of ageing on the in vivo elasticity of human skin. *Gerontology* 15 (2–3):121–139.

Green, M. A., L. E. Bilston and R. Sinkus. 2008. In vivo brain viscoelastic properties measured by magnetic resonance elastography. *NMR in Biomedicine* 21 (7):755–764.

Grove, G. L., J. Damia, M. Jo Grove and C. Zerweck. 2006. Suction chamber method for measurement of skin mechanics: The DermaLab. In *Handbook of Non-Invasive Methods and the Skin*, eds. J. Serup and G. Jemec. Boca Raton, FL: CRC, Chapter 68.

Gupta, H. S., S. Schratter, W. Tesch, P. Roschger, A. Berzlanovich, T. Schoeberl, K. Klaushofer, and P. Fratzl. 2005. Two different correlations between nanoindentation modulus and mineral content in the bone-cartilage interface. *Journal of Structural Biology* 149 (2):138–148.

Gupte, C. M., A. Smith, N. Jamieson, A. M. J. Bull, R. D. Thomas and A. A. Amis. 2002. Meniscofemoral ligaments-structural and material properties. *Journal of Biomechanics* 35 (12):1623–1629.

Habelitz, S., S. J. Marshall, G. W. Marshall and M. Balooch. 2001. Mechanical properties of human dental enamel on the nanometre scale. *Archives of Oral Biology* 46 (2):173–183.

Hall, T. J., P. E. Barbone, A. A. Oberai, J. F. Jiang, J. F. Dord, S. Goenezen and T. G. Fisher. 2011. Recent results in nonlinear strain and modulus imaging. *Current Medical Imaging Reviews* 7 (4):313–327.

Han, L. H., J. A. Noble and M. Burcher. 2003. A novel ultrasound indentation system for measuring biomechanical properties of *in vivo* soft tissue. *Ultrasound in Medicine and Biology* 29 (6):813–823.

Hansen, H. H. G., T. Idzenga and C. L. de Korte. 2012. Noninvasive vascular strain imaging: From methods to application. *Current Medical Imaging Reviews* 8 (1):37–45.

Haque, F. 2003. Application of nanoindentation to development of biomedical materials. *Surface Engineering* 19 (4):255–268.

Hattori, K., K. Uematsu, T. Matsumoto and H. Ohgushi. 2009. Mechanical effects of surgical procedures on osteochondral grafts elucidated by osmotic loading and real-time ultrasound. *Arthritis Research & Therapy* 11 (5):R134.

Hayes, W. C., G. Herrmann, L. F. Mockros and L. M. Keer. 1972. A mathematical analysis for indentation tests of articular cartilage. *Journal of Biomechanics* 5 (5):541–551.

Heikkila, J. and K. Hynynen. 2006. Investigation of optimal method for inducing harmonic motion in tissue using a linear ultrasound phased array – A simulation study. *Ultrasonic Imaging* 28 (2):97–113.

Hendriks, F. M., D. Brokken, C. W. J. Oomens and F. P. T. Baaijens. 2004. Influence of hydration and experimental length scale on the mechanical response of human skin in vivo, using optical coherence tomography. *Skin Research and Technology* 10 (4):231–241.

Hendriks, F. M., D. Brokken, C. W. J. Oomens, D. L. Bader and F. P. T. Baaijens. 2006. The relative contributions of different skin layers to the mechanical behavior of human skin in vivo using suction experiments. *Medical Engineering & Physics* 28 (3):259–266.

Hendriks, F. M., D. Brokken, J. T. W. M. van Eemeren, C. W. J. Oomens, F. P. T. Baaijens and J. B. A. M. Horsten. 2003. A numerical-experimental method to characterize the non-linear mechanical behaviour of human skin. *Skin Research and Technology* 9 (3):274–283.

Hillerton, J. E., S. E. Reynolds and J. F. V. Vincent. 1982. On the indentation hardness of insect cuticle. *Journal of Experimental Biology* 96:45–52.

Hochmuth, R. M. 2000. Micropipette aspiration of living cells. *Journal of Biomechanics* 33 (1):15–22.

Hollman, K. W., M. O'Donnell and T. N. Erpelding. 2007. Mapping elasticity in human lenses using bubble-based acoustic radiation force. *Experimental Eye Research* 85 (6):890–893.

Holzapfel, G. A. 2006. Determination of material models for arterial walls from uniaxial extension tests and histological structure. *Journal of Theoretical Biology* 238 (2):290–302.

Hon, Y. and A. K. C. Lam. 2013. Corneal deformation measurement using Scheimpflug noncontact tonometry. *Optometry and Vision Science* 90 (1): E1–E8.

Horgan, C. O. and G. Saccomandi. 2002. Constitutive modelling of rubber-like and biological materials with limiting chain extensibility. *Mathematics and Mechanics of Solids* 7 (4):353–371.

Horikawa, M., S. Ebihara, F. Sakai and M. Akiyama. 1993. Non-invasive measurement method for hardness in muscular tissues. *Medical & Biological Engineering & Computing* 31 (6):623–627.

Hoyt, K., B. Castaneda and K. J. Parker. 2008a. Two-dimensional sonoelastographic shear velocity imaging. *Ultrasound in Medicine and Biology* 34 (2):276–288.

Hoyt, K., T. Kneezel, B. Castaneda and K. J. Parker. 2008b. Quantitative sonoelastography for the in vivo assessment of skeletal muscle viscoelasticity. *Physics in Medicine and Biology* 53 (15):4063–4080.

Hoyt, K., K. J. Parker and D. J. Rubens. 2007. Real-time shear velocity imaging using sonoelastographic techniques. *Ultrasound in Medicine and Biology* 33 (7):1086–1097.

Hsu, C. C., W. C. Tsai, T. Y. Hsiao, F. Y. Tseng, Y. W. Shau, C. L. Wang and S. C. Lin. 2009. Diabetic effects on microchambers and macrochambers tissue properties in human heel pads. *Clinical Biomechanics* 24 (8):682–686.

Hsu, T. C., C. L. Wang, W. C. Tsai, J. K. Kuo and F. T. Tang. 1998. Comparison of the mechanical properties of the heel pad between young and elderly adults. *Archives of Physical Medicine and Rehabilitation* 79 (9):1101–1104.

Huang, Y. P. 2005. Ultrasonic assessment of post-radiotherapy fibrosis. MPhil thesis, Rehabilitation Engineering Center, Hong Kong Polytechnic University, Hung Hom, Hong Kong.

Huang, Y. P. 2013. Development of an arthroscopy-based water-jet ultrasound indentation system for the morphological, acoustic and mechanical assessment of articular cartilage degeneration. PhD, Interdisciplinary Division of Biomedical Engineering, Hong Kong Polytechnic University, Hung Hom, Hong Kong.

Huang, Y. P., S. Saarakkala, J. Toyras, L. K. Wang, J. S. Jurvelin and Y. P. Zheng. 2011a. Effects of optical beam angle on quantitative optical coherence tomography (OCT) in normal and surface degenerated bovine articular cartilage. *Physics in Medicine and Biology* 56 (2):491–509.

Huang, Y. P., S. Z. Wang, S. Saarakkala, and Y. P. Zheng. 2011b. Quantification of stiffness change in degenerated articular cartilage using optical coherence tomography-based air-jet indentation. *Connective Tissue Research* 52 (5):433–443.

Huang, Y. P. and Y. P. Zheng. 2009. Intravascular ultrasound (IVUS): A potential arthroscopic tool for quantitative assessment of articular cartilage. *The Open Biomedical Engineering Journal* 3:13–20.

Huang, Y. P. and Y. P. Zheng. 2013. Development of an arthroscopic ultrasound probe for assessment of articular cartilage degeneration. Paper read at *35th Annual International Conference of the IEEE Engineering in Medicine and Biology Society,* Osaka, Japan.

Huang, Y. P., Y. P. Zheng and S. F. Leung. 2005. Quasi-linear viscoelastic properties of fibrotic neck tissues obtained from ultrasound indentation tests *in vivo. Clinical Biomechanics* 20 (2):145–154.

Huang, Y. P., Y. P. Zheng, S. Z. Wang, Z. P. Chen, Q. H. Huang and Y. H. He. 2009. An optical coherence tomography (OCT)-based air jet indentation system for measuring the mechanical properties of soft tissues. *Measurement Science & Technology* 20 (1):015805.

Iatridis, J. C., L. A. Setton, R. J. Foster, B. A. Rawlins, M. Weidenbaum and V. Mow. 1998. Degeneration affects the anisotropic and nonlinear behaviors of human anulus fibrosus in compression. *Journal of Biomechanics* 31 (6):535–544.

Ichikawa, S., U. Motosugi, H. Morisaka, K. Sano, T. Ichikawa, A. Tatsumi, N. Enomoto, M. Matsuda, H. Fujii and H. Onishi. 2015. Comparison of the diagnostic accuracies of magnetic resonance elastography and transient elastography for hepatic fibrosis. *Magnetic Resonance Imaging* 33 (1):26–30.

Ingber, D. E. 2003. Mechanobiology and diseases of mechanotransduction. *Annals of Medicine* 35 (8):564–577.

Izatt, J. A., M. D. Kulkarni, H. W. Wang, K. Kobayashi and M. V. Sivak. 1996. Optical coherence tomography and microscopy in gastrointestinal tissues. *IEEE Journal of Selected Topics in Quantum Electronics* 2 (4):1017–1028.

Jachowicz, J., R. McMullen and D. Prettypaul. 2007. Indentometric analysis of in vivo skin and comparison with artificial skin models. *Skin Research and Technology* 13 (3):299–309.

Jalkanen, V. 2010. Hand-held resonance sensor for tissue stiffness measurements-a theoretical and experimental analysis. *Measurement Science & Technology* 21 (5):055801.

Jin, H. and J. L. Lewis. 2004. Determination of Poisson's ratio of articular cartilage by indentation using different-sized indenters. *Journal of Biomechanical Engineering – Transactions of the ASME* 126 (2):138–145.

Johnson, C. S., S. I. Mian, S. Moroi, D. Epstein, J. Izatt and N. A. Afshari. 2007. Role of corneal elasticity in damping of intraocular pressure. *Investigative Ophthalmology & Visual Science* 48 (6):2540–2544.

Johnson, K. L. 1985. Normal contact of elastic solids – Hertz theory. In *Contact Mechanics*, ed. K. L. Johnson. Cambridge, U.K.: Cambridge University Press, Chapter 4.

Joshi, M. D., J. K. Suh, T. Marui and S. L. Y. Woo. 1995. Interspecies variation of compressive biomechanical properties of the meniscus. *Journal of Biomedical Materials Research* 29 (7):823–828.

Julkunen, P., T. Harjula, J. Marjanen, H. J. Helminen and J. S. Jurvelin. 2009. Comparison of single-phase isotropic elastic and fibril-reinforced poroelastic models for indentation of rabbit articular cartilage. *Journal of Biomechanics* 42 (5):652–656.

Jurvelin, J. S., M. D. Buschmann and E. B. Hunziker. 1997. Optical and mechanical determination of Poisson's ratio of adult bovine humeral articular cartilage. *Journal of Biomechanics* 30 (3):235–241.

Jurvelin, J. S., T. Rasanen, P. Kolmonen and T. Lyyra. 1995. Comparison of optical, needle probe and ultrasonic techniques for the measurement of articular-cartilage thickness. *Journal of Biomechanics* 28 (2):231–235.

Kallel, F. and M. Bertrand. 1996. Tissue elasticity reconstruction using linear perturbation method. *IEEE Transactions on Medical Imaging* 15 (3):299–313.

Kallel, F., M. Bertrand and J. Ophir. 1996. Fundamental limitations on the contrast-transfer efficiency in elastography: An analytic study. *Ultrasound in Medicine and Biology* 22 (4):463–470.

Katz, S. M., D. H. Frank, G. R. Leopold and T. L. Wachtel. 1985. Objective measurement of hypertrophic burn scar: A preliminary study of tonometry and ultrasonography. *Annals of Plastic Surgery* 14 (2):121–127.

Kauer, M., V. Vuskovic, J. Dual, G. Szekely and M. Bajka. 2002. Inverse finite element characterization of soft tissues. *Medical Image Analysis* 6 (3):275–287.

Kaufman, J. D. and C. M. Klapperich. 2009. Surface detection errors cause overestimation of the modulus in nanoindentation on soft materials. *Journal of the Mechanical Behavior of Biomedical Materials* 2 (4):312–317.

Kaufman, J. D., G. J. Miller, E. F. Morgan and C. M. Klapperich. 2008. Time-dependent mechanical characterization of poly(2-hydroxyethyl methacrylate) hydrogels using nanoindentation and unconfined compression. *Journal of Materials Research* 23 (5):1472–1481.

Kawchuk, G. and W. Herzog. 1995. The reliability and accuracy of a standard method of tissue compliance assessment. *Journal of Manipulative and Physiological Therapeutics* 18 (5):298–301.

Kawchuk, G. N. and P. D. Elliott. 1998. Validation of displacement measurements obtained from ultrasonic images during indentation testing. *Ultrasound in Medicine and Biology* 24 (1):105–111.

Kawchuk, G. N., O. R. Fauvel and J. Dmowski. 2000. Ultrasonic quantification of osseous displacements resulting from skin surface indentation loading of bovine para-spinal tissue. *Clinical Biomechanics* 15 (4):228–233.

Kawchuk, G. N., O. R. Fauvel and J. Dmowski. 2001a. Ultrasonic indentation: A procedure for the noninvasive quantification of force-displacement properties of the lumbar spine. *Journal of Manipulative and Physiological Therapeutics* 24 (3):149–156.

Kawchuk, G. N., A. M. Kaigle, S. H. Holm, O. R. Fauvel, L. Ekstrom and T. Hansson. 2001b. The diagnostic performance of vertebral displacement measurements derived from ultrasonic indentation in an *in vivo* model of degenerative disc disease. *Spine* 26 (12):1348–1355.

Kerdok, A. E., S. M. Cotin, M. P. Ottensmeyer, A. M. Galea, R. D. Howe and S. L. Dawson. 2003. Truth cube: Establishing physical standards for soft tissue simulation. *Medical Image Analysis* 7 (3):283–291.

Kim, J. and S. Baek. 2011. Circumferential variations of mechanical behavior of the porcine thoracic aorta during the inflation test. *Journal of Biomechanics* 44 (10):1941–1947.

Kinney, J. H., S. J. Marshall and G. W. Marshall. 2003. The mechanical properties of human dentin: A critical review and re-evaluation of the dental literature. *Critical Reviews in Oral Biology & Medicine* 14 (1):13–29.

Kirk, E. and S. A. Kvorning. 1949. Quantitative measurements of the elastic properties of the skin and subcutaneous tissue in young and old individuals. *Journal of Gerontology* 4 (4):273–284.

Kirk, J. E. and M. Chieffi. 1962. Variation with age in elasticity of skin and subcutaneous tissue in human individuals. *Journal of Gerontology* 17:373–380.

Kiviranta, P., E. Lammentausta, J. Toyras, I. Kiviranta and J. S. Jurvelin. 2008. Indentation diagnostics of cartilage degeneration. *Osteoarthritis and Cartilage* 16 (7):796–804.

Kiviranta, P., E. Lammentausta, J. Toyras, H. J. Nieminen, P. Julkunen, I. Kiviranta and J. S. Jurvelin. 2009. Differences in acoustic properties of intact and degenerated human patellar cartilage during compression. *Ultrasound in Medicine and Biology* 35 (8):1367–1375.

Klaesner, J. W., P. K. Commean, M. K. Hastings, D. Q. Zou and M. J. Mueller. 2001. Accuracy and reliability testing of a portable soft tissue indentor. *IEEE Transactions on Neural Systems and Rehabilitation Engineering* 9 (2):232–240.

Klaesner, J. W., M. K. Hastings, D. Q. Zou, C. Lewis and M. J. Mueller. 2002. Plantar tissue stiffness in patients with diabetes mellitus and peripheral neuropathy. *Archives of Physical Medicine and Rehabilitation* 83 (12):1796–1801.

Klauser, A. S., H. Miyamoto, R. Bellmann-Weiler, G. M. Feuchtner, M. C. Wick and W. R. Jaschke. 2014. Sonoelastography: Musculoskeletal applications. *Radiology* 272 (3):622–633.

Knecht, S., B. Vanwanseele and E. Stussi. 2006. A review on the mechanical quality of articular cartilage – Implications for the diagnosis of osteoarthritis. *Clinical Biomechanics* 21 (10):999–1012.

Ko, M. W. L., L. K. K. Leung and D. C. C. Lam. 2014. Comparative study of corneal tangent elastic modulus measurement using corneal indentation device. *Medical Engineering & Physics* 36 (9):1115–1121.

Ko, M. W. L., L. K. K. Leung, D. C. C. Lam and C. K. S. Leung. 2013. Characterization of corneal tangent modulus in vivo. *Acta Ophthalmologica* 91 (4):E263–E269.

Kollias, G. E., K. S. Stamatelopoulos, T. G. Papaioannou, N. A. Zakopoulos, M. Alevizaki, G. P. Alexopoulos, D. A. Kontoyannis et al. 2009. Diurnal variation of endothelial function and arterial stiffness in hypertension. *Journal of Human Hypertension* 23 (9):597–604.

Komi, P. V., A. Belli, V. Huttunen, R. Bonnefoy, A. Geyssant and J. R. Lacour. 1996. Optic fibre as a transducer of tendomuscular forces. *European Journal of Applied Physiology and Occupational Physiology* 72 (3):278–80.

Konofagou, E. E., T. P. Harrigan, J. Ophir and T. A. Krouskop. 2001. Poroelastography: Imaging the poroelastic properties of tissues. *Ultrasound in Medicine and Biology* 27 (10):1387–1397.

Konofagou, E. E. and K. Hynynen. 2003. Localized harmonic motion imaging: Theory, simulations and experiments. *Ultrasound in Medicine and Biology* 29 (10):1405–1413.

Konofagou, E. E., C. Maleke and J. Vappou. 2012. Harmonic motion imaging (HMI) for tumor imaging and treatment monitoring. *Current Medical Imaging Reviews* 8 (1):16–26.

Konofagou, E. and J. Ophir. 1998. A new elastographic method for estimation and imaging of lateral displacements, lateral strains, corrected axial strains and Poisson's ratios in tissues. *Ultrasound in Medicine and Biology* 24 (8):1183–1199.

Korhonen, R. K., M. S. Laasanen, J. Toyras, J. Rieppo, J. Hirvonen, H. J. Helminen and J. S. Jurvelin. 2002. Comparison of the equilibrium response of articular cartilage in unconfined compression, confined compression and indentation. *Journal of Biomechanics* 35 (7):903–909.

Krouskop, T. A., D. R. Dougherty and F. S. Vinson. 1987. A pulsed Doppler ultrasonic system for making noninvasive measurements of the mechanical properties of soft tissue. *Journal of Rehabilitation Research and Development* 24 (2):1–8.

Krouskop, T. A., T. M. Wheeler, F. Kallel, B. S. Garra and T. Hall. 1998. Elastic moduli of breast and prostate tissues under compression. *Ultrasonic Imaging* 20 (4):260–274.

Kruse, S. A., G. H. Rose, K. J. Glaser, A. Manduca, J. P. Felmlee, C. R. Jack and R. L. Ehman. 2008. Magnetic resonance elastography of the brain. *Neuroimage* 39 (1):231–237.

Kudo, M., T. Shiina, F. Moriyasu, H. Iijima, R. Tateishi, N. Yada, K. Fujimoto et al. 2013. JSUM ultrasound elastography practice guidelines: Liver. *Journal of Medical Ultrasonics* 40 (4):325–357.

Kuei, S. C., W. M. Lai and V. C. Mow. 1978. A biphasic rheological model of articular cartilage. In *Advances in Bioengineering*, eds. R. C. Eberhardt and A. H. Burstein. New York: Americal Society of Mechanical Engineers.

Kuimova, M. K., S. W. Botchway, A. W. Parker, M. Balaz, H. A. Collins, H. L. Anderson, K. Suhling and P. R. Ogilby. 2009. Imaging intracellular viscosity of a single cell during photoinduced cell death. *Nature Chemistry* 1 (1):69–73.

Kusaka, K., Y. Harihara, G. Torzilli, K. Kubota, T. Takayama, M. Makuuchi, M. Mori and S. Omata. 2000. Objective evaluation of liver consistency to estimate hepatic fibrosis and functional reserve for hepatectomy. *Journal of the American College of Surgeons* 191 (1):47–53.

Kwan, R. L. C., Y. P. Zheng and G. L. Y. Cheing. 2010. The effect of aging on the biomechanical properties of plantar soft tissues. *Clinical Biomechanics* 25 (6):601–605.

Kydd, W. L., C. H. Daly and D. Nansen. 1974. Variation in the response to mechanical stress of human soft tissues as related to age. *Journal of Prosthetic Dentistry* 32 (5):493–500.

Laasanen, M. S., J. Toyras, J. Hirvonen, S. Saarakkala, R. K. Korhonen, M. T. Nieminen, I. Kiviranta and J. S. Jurvelin. 2002. Novel mechano-acoustic technique and instrument for diagnosis of cartilage degeneration. *Physiological Measurement* 23 (3):491–503.

Lau, J. C. M., C. W. P. Li-Tsang and Y. P. Zheng. 2005. Application of tissue ultrasound palpation system (TUPS) in objective scar evaluation. *Burns* 31 (4):445–452.

Lau, W. and D. Pye. 2011. Changes in corneal biomechanics and applanation tonometry with induced corneal swelling. *Investigative Ophthalmology & Visual Science* 52 (6):3207–3214.

Ledoux, W. R. and J. J. Blevins. 2007. The compressive material properties of the plantar soft tissue. *Journal of Biomechanics* 40 (13):2975–2981.

Lee, S. J., J. Sun, J. J. Flint, S. Guo, H. K. Xie, M. A. King and M. Sarntinoranont. 2011. Optically based-indentation technique for acute rat brain tissue slices and thin biomaterials. *Journal of Biomedical Materials Research Part B-Applied Biomaterials* 97B (1):84–95.

Legrice, I. J., Y. Takayama and J. W. Covell. 1995. Transverse shear along myocardial cleavage planes provides a mechanism for normal systolic wall thickening. *Circulation Research* 77 (1):182–193.

Leonard, C. T., J. U. Stephens and S. L. Stroppel. 2001. Assessing the spastic condition of individuals with upper motoneuron involvement: Validity of the myotonometer. *Archives of Physical Medicine and Rehabilitation* 82 (10):1416–1420.

Leonard, C. T. and E. L. Mikhailenok, 2000. Apparatus for measuring muscle tone, UPO, No. US6063044 A.

Lerner, R. M., S. R. Huang and K. J. Parker. 1990. Sonoelasticity images derived from ultrasound signals in mechanically vibrated tissues. *Ultrasound in Medicine and Biology* 16 (3):231–239.

Lerner, R. M. and K. J. Parker. 1987. Sonoelasticity images derived from ultrasound signals in mechanically vibrated targets. In *Seventh European Communities Workshop*, Nijmegen, the Netherlands.

Lerner, R. M., K. J. Parker, J. Holen, R. Gramiak and R. C. Waag. 1988. Sonoelasticity: Medical elasticity images derived from ultrasound signals in mechanically vibrated targets. *Acoustical Imaging* 16:317–327.

Leung, S. F., Y. P. Zheng, C. Y. K. Choi, S. S. S. Mak, S. K. W. Chiu, B. Zee and A. F. T. Mak. 2002. Quantitative measurement of post-irradiation neck fibrosis based on the Young modulus – Description of a new method and clinical results. *Cancer* 95 (3):656–662.

Lewis, G. and J. S. Nyman. 2008. The use of nanoindentation for characterizing the properties of mineralized hard tissues: State-of-the art review. *Journal of Biomedical Materials Research Part B – Applied Biomaterials* 87B (1):286–301.

Lewis, H. E., J. Mayer and A. A. Pandiscio. 1965. Recording skinfold calipers for the determination of subcutaneous edema. *Journal of Laboratory and Clinical Medicine* 66 (1):154–160.

Li, C. H., G. Guan, Z. Huang, M. Johnstone and R. K. Wang. 2012a. Noncontact all-optical measurement of corneal elasticity. *Optics Letters* 37 (10):1625–1627.

Li, C., L. A. Pruitt and K. B. King. 2006. Nanoindentation differentiates tissue-scale functional properties of native articular cartilage. *Journal of Biomedical Materials Research Part A* 78A (4):729–738.

Li, J., Y. Cui, M. Kadour and J. A. Noble. 2008a. Elasticity reconstruction from displacement and confidence measures of a multi-compressed ultrasound RF sequence. *IEEE Transactions on Ultrasonics Ferroelectrics and Frequency Control* 55 (2):319–326.

Li, J. W. 2009. The effects of menstrual cycle, site and individual variation on breast elasticity and thickness. MPhil, Health Technology and Informatics, Hong Kong Polytechnic University, Hung Hom, Hong Kong.

Li, J. W., S. T. Chan, Y. P. Huang and Y. P. Zheng. 2008b. Menstrual cycle dependences of *in vivo* breast elasticity measured using ultrasound indentation. In *The Seventh International Conference on the Ultrasonic Measurement and Imaging of Tissue Elasticity*, Austin, TX.

Li, J. W., S. T. Chan, Y. P. Huang and Y. P. Zheng. 2009. Menstrual cycle, site and individual dependences of breast elasticity measured *in vivo* using ultrasound indentation. In *The Eighth International Conference on the Ultrasonic Measurement and Imaging of Tissue Elasticity*, Vlissingen, Zeeland, the Netherlands.

Li, P., R. Reif, Z. W. Zhi, E. Martin, T. T. Shen, M. Johnstone and R. K. K. Wang. 2012b. Phase-sensitive optical coherence tomography characterization of pulse-induced trabecular meshwork displacement in ex vivo nonhuman primate eyes. *Journal of Biomedical Optics* 17 (7):076026.

Li, X. F., L. H. Huang, G. X. Zhang and T. Freiheit. 2005. Performance comparison of algorithms for estimating the strain of soft tissue from ultrasound. *Instrumentation Science & Technology* 33 (6):673–689.

Li, X. F., G. L. Wang, L. H. Huang and G. X. Zhang. 2006. Young's modulus extraction methods for soft tissue from ultrasound measurement system. *Instrumentation Science & Technology* 34 (4):393–404.

Li, X. F., S. Ying and T. Freiheit. 2004. A portable measurement instrument for soft tissue mechanical properties. *Instrumentation Science & Technology* 32 (6):611–626.

Liang, X., V. Crecea and S. A. Boppart. 2010. Dynamic optical coherence elastography: A review. *Journal of Innovative Optical Health Sciences* 3 (4):221–233.

Lim, C. T., E. H. Zhou, A. Li, S. R. K. Vedula and H. X. Fu. 2006. Experimental techniques for single cell and single molecule biomechanics. *Materials Science & Engineering C-Biomimetic and Supramolecular Systems* 26 (8):1278–1288.

Lin, D. C. and F. Horkay. 2008. Nanomechanics of polymer gels and biological tissues: A critical review of analytical approaches in the Hertzian regime and beyond. *Soft Matter* 4 (4):669–682.

Lindahl, O. A., C. E. Constantinou, A. Eklund, Y. Murayama, P. Hallberg and S. Omata. 2009. Tactile resonance sensors in medicine. *Journal of Medical Engineering & Technology* 33 (4):263–273.

Ling, H. Y., P. C. Choi, Y. P. Zheng and K. T. Lau. 2007a. Extraction of mechanical properties of foot plantar tissues using ultrasound indentation associated with genetic algorithm. *Journal of Materials Science – Materials in Medicine* 18 (8):1579–1586.

Ling, H. Y., Y. P. Zheng and S. G. Patil. 2007b. Strain dependence of ultrasound speed in bovine articular cartilage under compression *in vitro*. *Ultrasound in Medicine & Biology* 33 (10):1599–1608.

Lister, K., Z. Gao and J. P. Desai. 2011. Development of in vivo constitutive models for liver: Application to surgical simulation. *Annals of Biomedical Engineering* 39 (3):1060–1073.

Litwiller, D. V., Y. K. Mariappan and R. L. Ehman. 2012. Magnetic resonance elastography. *Current Medical Imaging Reviews* 8 (1):46–55.

Liu, H. B., J. C. Li, X. J. Song, L. D. Seneviratne and K. Althoefer. 2011. Rolling indentation probe for tissue abnormality identification during minimally invasive surgery. *IEEE Transactions on Robotics* 27 (3):450–460.

Liu, J. and C. J. Roberts. 2005. Influence of corneal biomechanical properties on intraocular pressure measurement – Quantitative analysis. *Journal of Cataract and Refractive Surgery* 31 (1):146–155.

Lorenzen, J., R. Sinkus, M. Biesterfeldt and G. Adam. 2003. Menstrual-cycle dependence of breast parenchyma elasticity: Estimation with magnetic resonance elastography of breast tissue during the menstrual cycle. *Investigative Radiology* 38 (4):236–240.

Lu, M. H. 2007. Development of a noncontact ultrasound indentation system for measuring tissue material properties using water jet. PhD thesis, Health Technology and Informatics, Hong Kong Polytechnic University, Hung Hom, Hong Kong.

Lu, M. H., W. N. Yu, Q. H. Huang, Y. P. Huang and Y. P. Zheng. 2009c. A hand-held indentation system for the assessment of mechanical properties of soft tissues in vivo. *IEEE Transactions on Instrumentation and Measurement* 58 (9):3079–3085.

Lu, M. H. and Y. P. Zheng. 2004. Indentation test of soft tissues with curved substrates: A finite element study. *Medical & Biological Engineering & Computing* 42 (4):535–540.

Lu, M. H., Y. P. Zheng and Q. H. Huang. 2005. A novel noncontact ultrasound indentation system for measurement of tissue material properties using water jet compression. *Ultrasound in Medicine and Biology* 31 (6):817–826.

Lu, M. H., Y. P. Zheng and Q. H. Huang. 2007. A novel method to obtain modulus image of soft tissues using ultrasound water jet indentation: A phantom study. *IEEE Transactions on Biomedical Engineering* 54 (1):114–121.

Lu, M. H., Y. P. Zheng, Q. H. Huang, C. Ling, Q. Wang, L. Bridal, L. Qin and A. Mak. 2009a. Noncontact evaluation of articular cartilage degeneration using a novel ultrasound water jet indentation system. *Annals of Biomedical Engineering* 37 (1):164–175.

Lu, M. H., R. Mao, Y. Lu, Z. Liu, T. F. Wang and S. P. Chen. 2012. Quantitative imaging of Young's modulus of soft tissues from ultrasound water jet indentation: A finite element study. *Computational and Mathematical Methods in Medicine* 2012:979847.

Lu, M. H., Y. P. Zheng, H. B. Lu, Q. H. Huang and L. Qin. 2009b. Evaluation of bone-tendon junction healing using water jet ultrasound indentation method. *Ultrasound in Medicine & Biology* 35 (11):1783–1793.

Lu, Y., K. H. Parker and W. Wang. 2006. Effects of osmotic pressure in the extracellular matrix on tissue deformation. *Philosophical Transactions of the Royal Society A – Mathematical Physical and Engineering Sciences* 364 (1843):1407–1422.

Luce, D. A. 2005. Determining in vivo biomechanical properties of the cornea with an ocular response analyzer. *Journal of Cataract and Refractive Surgery* 31 (1):156–162.

Lye, I., D. W. Edgar, F. M. Wood and S. Carroll. 2006. Tissue tonometry is a simple, objective measure for pliability of burn scar: Is it reliable? *Journal of Burn Care & Research* 27 (1):82–85.

Lyyra, T., J. Jurvelin, P. Pitkanen, U. Vaatainen and I. Kiviranta. 1995. Indentation instrument for the measurement of cartilage stiffness under arthroscopic control. *Medical Engineering & Physics* 17 (5):395–399.

Lyyra, T., I. Kiviranta, U. Vaatainen, H. J. Helminen and J. S. Jurvelin. 1999. In vivo characterization of indentation stiffness of articular cartilage in the normal human knee. *Journal of Biomedical Materials Research* 48 (4):482–487.

Mace, E., I. Cohen, G. Montaldo, R. Miles, M. Fink and M. Tanter. 2011. In vivo mapping of brain elasticity in small animals using shear wave imaging. *IEEE Transactions on Medical Imaging* 30 (3):550–558.

Mak, A. F., W. M. Lai and V. C. Mow. 1987. Biphasic indentation of articular cartilage. I. Theoretical analysis. *Journal of Biomechanics* 20:703–714.

Mak, A. F. T., G. H. W. Liu and S. Y. Lee. 1994. Biomechanical assessment of below-knee residual limb tissue. *Journal of Rehabilitation Research and Development* 31 (3):188–198.

Mak, T. M. 2013. Liver fibrosis assessment using transient elastography guided with real-time B-mode ultrasound imaging: A feasibility study. MPhil, Interdisciplinary Division of Biomedical Engineering, Hong Kong Polytechnic University, Hung Hom, Hong Kong.

Mak, T. M., Y. P. Huang and Y. P. Zheng. 2013. Liver fibrosis assessment using transient elastography guided with real-time B-mode ultrasound imaging: A feasbility study. *Ultrasound in Medicine and Biology* 39 (6):956–966.

Makhsous, M., G. Venkatasubramanian, A. Chawla, Y. Pathak, M. Priebe, W. Z. Rymer and F. Lin. 2008. Investigation of soft-tissue stiffness alteration in denervated human tissue using an ultrasound indentation system. *Journal of Spinal Cord Medicine* 31 (1):88–96.

Maleke, C. and E. E. Konofagou. 2008. Harmonic motion imaging for focused ultrasound (HMIFU): A fully integrated technique for sonication and monitoring of thermal ablation in tissues. *Physics in Medicine and Biology* 53 (6):1773–1793.

Maleke, C., M. Pernot and E. E. Konofagou. 2006. Single-element focused ultrasound transducer method for harmonic motion imaging. *Ultrasonic Imaging* 28 (3):144–158.

Manapuram, R. K., S. R. Aglyamov, F. M. Monediado, M. Mashiatulla, J. S. Li, S. Y. Emelianov and K. V. Larin. 2012. In vivo estimation of elastic wave parameters using phase-stabilized swept source optical coherence elastography. *Journal of Biomedical Optics* 17 (10):100501.

Margulies, S. S., L. E. Thibault and T. A. Gennarelli. 1990. Physical model simulations of brain injury in the primate. *Journal of Biomechanics* 23 (8):823-826.

Mariappan, Y. K., K. J. Glaser and R. L. Ehman. 2010. Magnetic resonance elastography: A review. *Clinical Anatomy* 23 (5):497–511.

Martins, P., R. M. N. Jorge and A. J. M. Ferreira. 2006. A comparative study of several material models for prediction of hyperelastic properties: Application to silicone-rubber and soft tissues. *Strain* 42 (3):135–147.

Mathiesen, O., L. Konradsen, S. Torp-Pedersen and U. Jorgensen. 2004. Ultrasonography and articular cartilage defects in the knee: An in vitro evaluation of the accuracy of cartilage thickness and defect size assessment. *Knee Surgery Sports Traumatology Arthroscopy* 12 (5):440–443.

Mazza, E., P. Grau, M. Hollenstein and M. Bajka. 2008. Constitutive modeling of human liver based on in vivo measurements. In *Proceedings of the Medical Image Computing and Computer-Assisted Intervention – Miccai 2008, Pt Ii*, eds. D. Metaxas, L. Axel, G. Fichtinger and G. Szekely. Berlin, Germany: Springer-Verlag.

Mazza, E., A. Nava, M. Bauer, R. Winter, M. Bajka and G. A. Holzapfel. 2006. Mechanical properties of the human uterine cervix: An in vivo study. *Medical Image Analysis* 10 (2):125–136.

Mazza, E., A. Nava, D. Halmloser, W. Jochum and M. Bajka. 2007. The mechanical response of human liver and its relation to histology: An *in vivo* study. *Medical Image Analysis* 11 (6):663–672.

McAleavey, S. and M. Menon. 2007. Direct estimation of shear modulus using spatially modulated acoustic radiation force impulses. In *2007 IEEE Ultrasonics Symposium Proceedings*, Vols. 1–6, pp. 558–561. New York: IEEE.

McDannold, N. and S. E. Maier. 2008. Magnetic resonance acoustic radiation force imaging. *Medical Physics* 35 (8):3748–3758.

McKee, C. T., J. A. Last, P. Russell and C. J. Murphy. 2011. Indentation versus tensile measurements of Young's modulus for soft biological tissues. *Tissue Engineering Part B-Reviews* 17 (3):155–164.

McKnight, A. L., J. L. Kugel, P. J. Rossman, A. Manduca, L. C. Hartmann and R. L. Ehman. 2002. MR elastography of breast cancer: Preliminary results. *American Journal of Roentgenology* 178 (6):1411–1417.

McLaughlin, J. and D. Renzi. 2006a. Shear wave speed recovery in transient elastography and supersonic imaging using propagating fronts. *Inverse Problems* 22 (2):681–706.

McLaughlin, J. and D. Renzi. 2006b. Using level set based inversion of arrival times to recover shear wave speed in transient elastography and supersonic imaging. *Inverse Problems* 22 (2):707–725.

McLaughlin, J., D. Renzi, K. Parker and Z. Wu. 2007. Shear wave speed recovery using moving interference patterns obtained in sonoelastography experiments. *Journal of the Acoustical Society of America* 121 (4):2438–2446.

Merkel, P. A., N. P. Silliman, C. P. Denton, D. E. Furst, D. Khanna, P. Emery, V. M. Hsu et al. 2008. Validity, reliability, and feasibility of durometer measurements of scleroderma skin disease in a multicenter treatment trial. *Arthritis & Rheumatism-Arthritis Care & Research* 59 (5):699–705.

Miller, K. 2000. Biomechanics of soft tissues. *Medical Science Monitor* 6 (1):158–167.

Miller, K. and K. Chinzei. 1997. Constitutive modelling of brain tissue: Experiment and theory. *Journal of Biomechanics* 30 (11–12):1115–1121.

Miller, K. and K. Chinzei. 2002. Mechanical properties of brain tissue in tension. *Journal of Biomechanics* 35 (4):483–490.

Miller-Young, J. E., N. A. Duncan and G. Baroud. 2002. Material properties of the human calcaneal fat pad in compression: Experiment and theory. *Journal of Biomechanics* 35 (12):1523–1531.

Mirnajafi, A., J. M. Raymer, L. R. McClure and M. S. Sacks. 2006. The flexural rigidity of the aortic valve leaflet in the commissural region. *Journal of Biomechanics* 39 (16):2966–2973.

Mirnajafi, A., J. Raymer, M. J. Scott and M. S. Sacks. 2005. The effects of collagen fiber orientation on the flexural properties of pericardial heterograft biomaterials. *Biomaterials* 26 (7):795–804.

Mirnajafi, A., A. Moseley and N. Piller. 2004. A new technique for measuring skin changes of patients with chronic postmastectomy lymphedema. *Lymphatic Research and Biology* 2 (2):82–85.

Misra, S., K. T. Ramesh and A. M. Okamura. 2008. Modeling of tool-tissue interactions for computer-based surgical simulation: A literature review. *Presence-Teleoperators and Virtual Environments* 17 (5):463–491.

Mitchison, J. M. and M. M. Swann. 1954. The mechanical properties of the cell surface: I. The cell elastimeter. *Journal of Experimental Biology* 31 (3):443–460.

Mitri, F. G., B. J. Davis, M. W. Urban, A. Alizad, J. F. Greenleaf, G. H. Lischer, T. M. Wilson and M. Fatemi. 2009. Vibro-acoustography imaging of permanent prostate brachytherapy seeds in an excised human prostate – Preliminary results and technical feasibility. *Ultrasonics* 49 (3):389–394.

Mitri, F. G., M. W. Urban, M. Fatemi and J. F. Greenleaf. 2011. Shear wave dispersion ultrasonic vibrometry for measuring prostate shear stiffness and viscosity: An in vitro pilot study. *IEEE Transactions on Biomedical Engineering* 58 (2):235–242.

Miyaji, K., A. Furuse, J. Nakajima, T. Kohno, T. Ohtsuka, K. Yagyu, T. Oka and S. Omata. 1997. The stiffness of lymph nodes containing lung carcinoma metastases – A new diagnostic parameter measured by a tactile sensor. *Cancer* 80 (10):1920–1925.

Monson, K. L., W. Goldsmith, N. M. Barbaro and G. T. Manley. 2003. Axial mechanical properties of fresh human cerebral blood vessels. *Journal of Biomechanical Engineering – Transactions of the ASME* 125 (2):288–294.

Mow, V. C., M. C. Gibbs, W. M. Lai, W. B. Zhu and K. A. Athanasiou. 1989. Biphasic indentation of articular cartilage. II. A numerical algorithm and an experimental study. *Journal of Biomechanics* 22:853–861.

Mow, V. C., S. C. Kuei, W. M. Lai and C. G. Armstrong. 1980. Biphasic creep and stress-relaxation of articular cartilage in compression – Theory and experiments. *Journal of Biomechanical Engineering – Transactions of the ASME* 102 (1):73–84.

Muller, M., J. L. Gennisson, T. Deffieux, M. Tanter and M. Fink. 2009. Quantitative viscoelasticity mapping of human liver using supersonic shear imaging: Preliminary *in vivo* feasibility study. *Ultrasound in Medicine and Biology* 35 (2):219–229.

Murayama, Y., C. E. Constantinou and S. Omata. 2004. Micro-mechanical sensing platform for the characterization of the elastic properties of the ovum via uniaxial measurement. *Journal of Biomechanics* 37 (1):67–72.

Murayama, Y., M. Haruta, Y. Hatakeyama, T. Shiina, H. Sakuma, S. Takenoshita, S. Omata and C. E. Constantinou. 2008. Development of a new instrument for examination of stiffness in the breast using haptic sensor technology. *Sensors and Actuators A – Physical* 143 (2):430–438.

Muthupillai, R., D. J. Lomas, P. J. Rossman, J. F. Greenleaf, A. Manduca and R. L. Ehman. 1995. Magnetic resonance elastography by direct visualization of propagating acoustic strain waves. *Science* 269 (5232):1854–1857.

Myers, S. L., K. Dines, D. A. Brandt, K. D. Brandt and M. E. Albrecht. 1995. Experimental assessment by high-frequency ultrasound of articular cartilage thickness and osteoarthritic changes. *Journal of Rheumatology* 22 (1):109–116.

Nakashima, K., T. Shiina, M. Sakurai, K. Enokido, T. Endo, H. Tsunoda, E. Takada, T. Umemoto and E. Ueno. 2013. JSUM ultrasound elastography practice guidelines: Breast. *Journal of Medical Ultrasonics* 40 (4):359–391.

Narmoneva, D. A., H. S. Cheung, J. Y. Wang, D. S. Howell and L. A. Setton. 2002. Altered swelling behavior of femoral cartilage following joint immobilization in a canine model. *Journal of Orthopaedic Research* 20 (1):83–91.

Narmoneva, D. A., J. Y. Wang and L. A. Setton. 1999. Nonuniform swelling-induced residual strains in articular cartilage. *Journal of Biomechanics* 32 (4):401–408.

Narmoneva, D. A., J. Y. Wang and L. A. Setton. 2001. A noncontacting method for material property determination for articular cartilage from osmotic loading. *Biophysical Journal* 81 (6):3066–3076.

Nava, A., E. Mazza, M. Furrer, P. Villiger and W. H. Reinhart. 2008. In vivo mechanical characterization of human liver. *Medical Image Analysis* 12 (2):203–216.

Nava, A., E. Mazza, F. Kleinermann, N. J. Avis, J. McClure and M. Bajka. 2004. Evaluation of the mechanical properties of human liver and kidney through aspiration experiments. *Technology and Health Care* 12 (3):269–280.

Ng, C. O. Y., G. Y. F. Ng, E. K. N. See and M. C. P. Leung. 2003. Therapeutic ultrasound improves strength of Achilles tendon repair in rats. *Ultrasound in Medicine and Biology* 29 (10):1501–1506.

Nguyen, T. M., B. Arnal, S. Z. Song, Z. H. Huang, R. K. Wang and M. O'Donnell. 2015. Shear wave elastography using amplitude-modulated acoustic radiation force and phase-sensitive optical coherence tomography. *Journal of Biomedical Optics* 20 (1):016001.

Nguyen, T. M., M. Couade, J. Bercoff and M. Tanter. 2011. Assessment of viscous and elastic properties of sub-wavelength layered soft tissues using shear wave spectroscopy: Theoretical framework and in vitro experimental validation. *IEEE Transactions on Ultrasonics Ferroelectrics and Frequency Control* 58 (11):2305–2315.

Nicolson, M. 1993. The art of diagnosis: Medicine and the five senses. In *Companion Encyclopedia of the History of Medicine*, eds. W. F. Bynum and R. Porter. London, U.K.: Routledge.

Nicosia, M. A. 2007. A theoretical framework to analyze bend testing of soft tissue. *Journal of Biomechanical Engineering – Transactions of the ASME* 129 (1):117–120.

Niederauer, G. G., G. M. Niederauer, L. C. Cullen, K. A. Athanasiou, J. B. Thomas and M. Q. Niederauer. 2004. Correlation of cartilage stiffness to thickness and level of degeneration using a handheld indentation probe. *Annals of Biomedical Engineering* 32 (3):352–359.

Nieminen, H. J., P. Julkunen, J. Toyras and J. S. Jurvelin. 2007. Ultrasound speed in articular cartilage under mechanical compression. *Ultrasound in Medicine and Biology* 33 (11):1755–1766.

Nieminen, H. J., Y. P. Zheng, S. Saarakkala, Q. Wang, J. Toyras, Y. P. Huang and J. Jurvelin. 2009. Quantitative assessment of articular cartilage using high-frequency ultrasound: Research findings and diagnostic prospects. *Critical Reviews in Biomedical Engineering* 37 (6):461–94.

Nightingale, K. 2011. Acoustic radiation force impulse (ARFI) imaging: A review. *Current Medical Imaging Reviews* 7 (4):328–339.

Nightingale, K., S. McAleavey and G. Trahey. 2003. Shear-wave generation using acoustic radiation force: In vivo and ex vivo results. *Ultrasound in Medicine and Biology* 29 (12):1715–1723.

Nightingale, K., R. Nightingale, M. Palmeri and G. Trahey. 1999. Finite element analysis of radiation force induced tissue motion with experimental validation. In *IEEE Ultrasonics Symposium Proceedings,* Vols. 1 and 2, eds. S. C. Schneider, M. Levy and B. R. McAvoy. New York: IEEE.

Nightingale, K., M. S. Soo, R. Nightingale and G. Trahey. 2002. Acoustic radiation force impulse imaging: In vivo demonstration of clinical feasibility. *Ultrasound in Medicine and Biology* 28 (2):227–235.

Nii, K., N. Tagawa, K. Okubo and S. Yagi. 2013. Finite element method study for generating shear wave by mode conversion of longitudinal wave at elasticity boundary in a living body. *Japanese Journal of Applied Physics* 52 (7):07FH23.

Nordez, A. and F. Hug. 2010. Muscle shear elastic modulus measured using supersonic shear imaging is highly related to muscle activity level. *Journal of Applied Physiology* 108 (5):1389–1394.

Nyce, D. S. 2003. *Linear Position Sensors: Theory and Application*. Hoboken, NJ: John Wiley & Sons Inc.

O'goshi, K. I. 2006. Suction chamber method for measurement of skin mechanics: The Cutometer'. In *Handbook of Non-Invasive Methods and the Skin*, eds. J. Serup and G. Jemec. Boca Raton, FL: CRC, Chapter 66.

O'Hagan, J. J. and A. Samani. 2009. Measurement of the hyperelastic properties of 44 pathological ex vivo breast tissue samples. *Physics in Medicine and Biology* 54 (8):2557–2569.

Oberai, A. A., N. H. Gokhale, S. Goenezen, P. E. Barbone, T. J. Hall, A. M. Sommer and J. F. Jiang. 2009. Linear and nonlinear elasticity imaging of soft tissue in vivo: Demonstration of feasibility. *Physics in Medicine and Biology* 54 (5):1191–1207.

Ohshima, H., S. Kinoshita, M. Oyobikawa, M. Futagawa, H. Takiwaki, A. Ishiko and H. Kanto. 2013. Use of Cutometer area parameters in evaluating age-related changes in the skin elasticity of the cheek. *Skin Research and Technology* 19 (1):E238–E242.

Oliver, W. C. and G. M. Pharr. 1992. An improved technique for determining hardness and elastic modulus using load and displacement sensing indentation experiments. *Journal of Materials Research* 7 (6):1564–1583.

Oliver, W. C. and G. M. Pharr. 2004. Measurement of hardness and elastic modulus by instrumented indentation: Advances in understanding and refinements to methodology. *Journal of Materials Research* 19 (1):3–20.

Omata, S., Y. Murayama and C. E. Constantinou. 2004. Real time robotic tactile sensor system for the determination of the physical properties of biomaterials. *Sensors and Actuators A – Physical* 112 (2–3):278–285.

Omata, S. and Y. Terunuma. 1992. New tactile sensor like the human hand and its applications. *Sensors and Actuators A – Physical* 35 (1):9–15.

Ophir, J., I. Cespedes, B. Garra, H. Ponnekanti, Y. Huang and N. Maklad. 1996. Elastography: Ultrasonic imaging of tissue strain and elastic modulus *in vivo*. *European Journal of Ultrasound* 3 (1):49–70.

Ophir, J., I. Cespedes, H. Ponnekanti, Y. Yazdi and X. Li. 1991. Elastography – A quantitative method for imaging the elasticity of biological tissues. *Ultrasonic Imaging* 13 (2):111–134.

Ophir, J., F. Kallel, T. Varghese, M. Bertrand, I. Cespedes and H. Ponnekanti. 1997. Elastography: A systems approach. *International Journal of Imaging Systems and Technology* 8 (1):89–103.

Ophir, J., S. Srinivasan, R. Righetti and A. Thittai. 2011. Elastography: A decade of progress (2000–2010). *Current Medical Imaging Reviews* 7 (4):292–312.

Ortiz, D., D. Pinero, M. H. Shabayek, F. Arnalich-Montiel and J. L. Alio. 2007. Corneal biomechanical properties in normal, post-laser in situ keratomileusis, and keratoconic eyes. *Journal of Cataract and Refractive Surgery* 33 (8):1371–1375.

Oyen, M. L. 2005. Spherical indentation creep following ramp loading. *Journal of Materials Research* 20 (8):2094–2100.

Oyen, M. L. 2011a. *Handbook of Nanoindentation with Biological Applications.* Singapore: Pan Stanford.

Oyen, M. L. 2011b. Nanoindentation of biological and biomimetic materials. *Experimental Techniques* 1–15.

Ozolanta, I., G. Tetere, B. Purinya and V. Kasyanov. 1998. Changes in the mechanical properties, biochemical contents and wall structure of the human coronary arteries with age and sex. *Medical Engineering & Physics* 20 (7):523–533.

Pai, S. and W. R. Ledoux. 2011. The quasi-linear viscoelastic properties of diabetic and non-diabetic plantar soft tissue. *Annals of Biomedical Engineering* 39 (5):1517–1527.

Pallwein, L., M. Mitterberger, J. Gradl, F. Aigner, W. Horninger, H. Strasser, G. Bartsch, D. zur Nedden and F. Frauscher. 2007. Value of contrast-enhanced ultrasound and elastography in imaging of prostate cancer. *Current Opinion in Urology* 17 (1):39–47.

Palmeri, M. L., K. D. Frinkley and K. R. Nightingale. 2004. Experimental studies of the thermal effects associated with radiation force imaging of soft tissue. *Ultrasonic Imaging* 26 (2):100–114.

Palmeri, M. L. and K. R. Nightingale. 2004. On the thermal effects associated with radiation force imaging of soft tissue. *IEEE Transactions on Ultrasonics Ferroelectrics and Frequency Control* 51 (5):551–565.

Pan, L., L. Zan and F. S. Foster. 1998. Ultrasonic and viscoelastic properties of skin under transverse mechanical stress in vitro. *Ultrasound in Medicine and Biology* 24 (7):995–1007.

Pan, Y. T., Z. G. Li, T. Q. Xie and C. R. Chu. 2003. Hand-held arthroscopic optical coherence tomography for *in vivo* high-resolution imaging of articular cartilage. *Journal of Biomedical Optics* 8 (4):648–654.

Parker, K. J. 2011. The evolution of vibration sonoelastography. *Current Medical Imaging Reviews* 7 (4):283–291.

Parker, K. J., M. M. Doyley and D. J. Rubens. 2011. Imaging the elastic properties of tissue: The 20 year perspective. *Physics in Medicine and Biology* 56 (1):R1–R29.

Parker, K. J., S. R. Huang, R. A. Musulin and R. M. Lerner. 1990. Tissue response to mechanical vibrations for sonoelasticity imaging. *Ultrasound in Medicine and Biology* 16 (3):241–246.

Parker, K. J., L. S. Taylor, S. Gracewski and D. J. Rubens. 2005. A unified view of imaging the elastic properties of tissue. *Journal of the Acoustical Society of America* 117 (5):2705–2712.

Pathak, A. P., M. B. Silver-Thorn, C. A. Thierfelder and T. E. Prieto. 1998. A rate-controlled indentor for *in vivo* analysis of residual limb tissues. *IEEE Transactions on Rehabilitation Engineering* 6 (1):12–20.

Patil, S. G. 2005. Measurement of the sound speed in articular cartilage in vitro. MPhil, Rehabilitation Engineering Center, Hong Kong Polytechnic University, Hung Hom, Hong Kong.

Pavan, T. Z. and A. A. O. Carneiro. 2011. Change in strain image contrast with applied compression evaluated via finite elements analysis. Paper read at *2011 Pan American Health Care Exchanges (PAHCE)*, March 28–April 1, 2011.

Peck, S. M. and A. W. Glick. 1956. A new method for measuring the hardness of keratin. *Journal of the Society of Cosmetic Chemists* 7:530–540.

Pedrini, G., M. Gusev, S. Schedin and H. J. Tiziani. 2003. Pulsed digital holographic interferometry by using a flexible fiber endoscope. *Optics and Lasers in Engineering* 40 (5–6):487–499.

Qin, L., Y. P. Zheng, C. T. Leung, A. Mak, W. Y. Choy, and K. M. Chan. 2002. Ultrasound detection of trypsin-treated articular cartilage: Its association with cartilaginous proteoglycans assessed by histological and biochemical methods. *Journal of Bone and Mineral Metabolism* 20 (5):281–287.

Raghavan, K. R. and A. E. Yagle. 1994. Forward and inverse problems in elasticity imaging of soft tissues. *IEEE Transactions on Nuclear Science* 41 (4):1639–1648.

Rashid, B., M. Destrade and M. D. Gilchrist. 2012. Determination of friction coefficient in unconfined compression of brain tissue. *Journal of the Mechanical Behavior of Biomedical Materials* 14:163–171.

Rayleigh, L. 1902. XXXIV. On the pressure of vibrations. *Philosophical Magazine Series 6* 3 (15):338–346.

Revel, G. M., A. Scalise and L. Scalise. 2003. Measurement of stress-strain and vibrational properties of tendons. *Measurement Science & Technology* 14 (8):1427–1436.

Rho, J. Y., T. Y. Tsui and G. M. Pharr. 1997. Elastic properties of human cortical and trabecular lamellar bone measured by nanoindentation. *Biomaterials* 18 (20):1325–1330.

Richards, M. S., P. E. Barbone and A. A. Oberai. 2009. Quantitative three-dimensional elasticity imaging from quasi-static deformation: A phantom study. *Physics in Medicine and Biology* 54 (3):757–779.

Ringleb, S. I., S. F. Bensamoun, Q. S. Chen, A. Manduca, K. N. An and R. L. Ehman. 2007. Applications of magnetic resonance elastography to healthy and pathologic skeletal muscle. *Journal of Magnetic Resonance Imaging* 25 (2):301–309.

Roemhildt, M. L., K. M. Coughlin, G. D. Peura, B. C. Fleming and B. D. Beynnon. 2006. Material properties of articular cartilage in the rabbit tibial plateau. *Journal of Biomechanics* 39 (12):2331–2337.

Romano, C., P. Rubegni, G. De Aloe, E. Stanghellini, G. D'Ascenzo, L. Andreassi and M. Fimiani. 2003. Extracorporeal photochemotherapy in the treatment of eosinophilic fasciitis. *Journal of the European Academy of Dermatology and Venereology* 17 (1):10–13.

Rome, K. and P. Webb. 2000. Development of a clinical instrument to measure heel pad indentation. *Clinical Biomechanics* 15 (4):298–300.

Rome, K., P. Webb, A. Unsworth and I. Haslock. 2001. Heel pad stiffness in runners with plantar heel pain. *Clinical Biomechanics* 16 (10):901–905.

Rosales, P. and S. Marcos. 2009. Pentacam Scheimpflug quantitative imaging of the crystalline lens and intraocular lens. *Journal of Refractive Surgery* 25 (5):421–428.

Rouviere, O., M. Yin, M. A. Dresner, P. J. Rossman, L. J. Burgart, J. L. Fidler and R. L. Ehman. 2006. MR elastography of the liver: Preliminary results. *Radiology* 240 (2):440–448.

Roy, R., S. S. Kohles, V. Zaporojan, G. M. Peretti, M. A. Randolph, J. W. Xu and L. J. Bonassar. 2004. Analysis of bending behavior of native and engineered auricular and costal cartilage. *Journal of Biomedical Materials Research Part A* 68A (4):597–602.

Royston, T. J., Z. J. Dai, R. Chaunsali, Y. F. Liu, Y. Peng and R. L. Magin. 2011. Estimating material viscoelastic properties based on surface wave measurements: A comparison of techniques and modeling assumptions. *Journal of the Acoustical Society of America* 130 (6):4126–4138.

Ruland, R. T., C. J. Hogan, C. J. Randall, A. Richards,and S. M. Belkoff. 2008. Biomechanical comparison of ulnar collateral ligament reconstruction techniques. *American Journal of Sports Medicine* 36 (8):1565–1570.

Ryu, H. S., Y. H. Joo, S. O. Kim, K. C. Park and S. W. Youn. 2008. Influence of age and regional differences on skin elasticity as measured by the Cutometer (R). *Skin Research and Technology* 14 (3):354–358.

Saarakkala, S., M. S. Laasanen, J. S. Jurvelin, K. Torronen, M. J. Lammi, R. Lappalainen and J. Toyras. 2003. Ultrasound indentation of normal and spontaneously degenerated bovine articular cartilage. *Osteoarthritis and Cartilage* 11 (9):697–705.

Saarakkala, S., J. Toyras, J. Hirvonen, M. S. Laasanen, R. Lappalainen and J. S. Jurvelin. 2004. Ultrasonic quantitation of superficial degradation of articular cartilage. *Ultrasound in Medicine and Biology* 30 (6):783–792.

Saarakkala, S., S. Z. Wang, Y. P. Huang and Y. P. Zheng. 2009. Quantification of optical surface reflection and surface roughness of articular cartilage using optical coherence tomography. *Physics in Medicine and Biology* 54 (22):6837–6852.

Sack, I., B. Beierbach, U. Hamhaber, D. Klatt and A. Braun. 2008. Non-invasive measurement of brain viscoelasticity using magnetic resonance elastography. *NMR in Biomedicine* 21 (3):265–271.

Sack, I., K. Johrens, J. Wurfel and J. Braun. 2013. Structure-sensitive elastography: On the viscoelastic powerlaw behavior of in vivo human tissue in health and disease. *Soft Matter* 9 (24):5672–5680.

Samani, A. and D. Plewes. 2004. A method to measure the hyperelastic parameters of ex vivo breast tissue samples. *Physics in Medicine and Biology* 49 (18):4395–4405.

Samani, A., J. Zubovits and D. Plewes. 2007. Elastic moduli of normal and pathological human breast tissues: An inversion-technique-based investigation of 169 samples. *Physics in Medicine and Biology* 52 (6):1565–1576.

Samur, E., M. Sedef, C. Basdogan, L. Avtan and O. Duzgun. 2007. A robotic indenter for minimally invasive measurement and characterization of soft tissue response. *Medical Image Analysis* 11 (4):361–373.

Sanders, G. E. and D. A. Lawson. 1992. Stability of paraspinal tissue compliance in normal subjects. *Journal of Manipulative and Physiological Therapeutics* 15 (6):361–364.

Sandrin, L., S. Catheline, M. Tanter, X. Hennequin and M. Fink. 1999. Time-resolved pulsed elastography with ultrafast ultrasonic imaging. *Ultrasonic Imaging* 21 (4):259–272.

Sandrin, L., B. Fourquet, J. M. Hasquenoph, S. Yon, C. Fournier, F. Mal, C. Christidis et al. 2003. Transient elastography: A new noninvasive method for assessment of hepatic fibrosis. *Ultrasound in Medicine and Biology* 29 (12):1705–1713.

Sandrin, L., M. Tanter, J. L. Gennisson, S. Catheline and M. Fink. 2002. Shear elasticity probe for soft tissues with 1-D transient elastography. *IEEE Transactions on Ultrasonics Ferroelectrics and Frequency Control* 49 (4):436–446.

Sarvazyan, A. and V. Egorov. 2012. Mechanical imaging – A technology for 3-D visualization and characterization of soft tissue abnormalities: A review. *Current Medical Imaging Reviews* 8 (1):64–73.

Sarvazyan, A., T. J. Hall, M. W. Urban, M. Fatemi, S. R. Aglyamov and B. S. Garra. 2011. An overview of elastography – An emerging branch of medical imaging. *Current Medical Imaging Reviews* 7 (4):255–282.

Sarvazyan, A. P., O. V. Rudenko and W. L. Nyborg. 2010. Biomedical applications of radiation force of ultrasound: Historical roots and physical basis. *Ultrasound in Medicine and Biology* 36 (9):1379–1394.

Sarvazyan, A. P., O. V. Rudenko, S. D. Swanson, J. B. Fowlkes and S. Y. Emelianov. 1998. Shear wave elasticity imaging: A new ultrasonic technology of medical diagnostics. *Ultrasound in Medicine and Biology* 24 (9):1419–1435.

Schade, H. 1912. Untersuchungen zur organfunction des bindegewebes. *Zeitschrift für die experimentelle Pathologie und Therapie* 11 (3):369–399.

Schiavone, P., F. Chassat, T. Boudou, E. Promayon, F. Valdivia and Y. Payan. 2009. In vivo measurement of human brain elasticity using a light aspiration device. *Medical Image Analysis* 13 (4):673–678.

Schiavone, P., E. Promayon and Y. Payan. 2010. LASTIC: A light aspiration device for *in vivo* soft TIssue characterization. In *Proceedings of the Biomedical Simulation,* eds. F. Bello and S. Cotin. Berlin, Germany: Springer-Verlag.

Schinagl, R. M., D. Gurskis, A. C. Chen and R. L. Sah. 1997. Depth-dependent confined compression modulus of full-thickness bovine articular cartilage. *Journal of Orthopaedic Research* 15 (4):499–506.

Schneider, S., A. Peipsi, M. Stokes, A. Knicker and V. Abeln. 2015. Feasibility of monitoring muscle health in microgravity environments using Myoton technology. *Medical & Biological Engineering & Computing* 53 (1):57–66.

Sebag, F., J. Vaillant-Lombard, J. Berbis, V. Griset, J. F. Henry, P. Petit and C. Oliver. 2010. Shear wave elastography: A new ultrasound imaging mode for the differential diagnosis of benign and malignant thyroid nodules. *Journal of Clinical Endocrinology & Metabolism* 95 (12):5281–5288.

Sebastiani, G. and A. Alberti. 2006. Non invasive fibrosis biomarkers reduce but not substitute the need for liver biopsy. *World Journal of Gastroenterology* 12 (23):3682–3694.

Setton, L. A., V. C. Mow and D. S. Howell. 1995. Mechanical behavior of articular cartilage in shear is altered by transection of the anterior ligament. *Journal of Orthopaedic Research* 13 (4):473–482.

Setton, L. A., H. Tohyama and V. C. Mow. 1998. Swelling and curling behaviors of articular cartilage. *Journal of Biomechanical Engineering – Transactions of the ASME* 120 (3):355–361.

Shadwick, R. E. and J. M. Gosline. January 1985. Mechanical properties of the octopus aorta. *Journal of Experimental Biology* 114:259–284.

Shao, J. Y., G. Xu and P. Guo. 2004. Quantifying cell-adhesion strength with micropipette manipulation: Principle and application. *Frontiers in Bioscience* 9:2183–2191.

Shen, J. J., M. Kalantari, J. Kovecses, J. Angeles and J. Dargahi. 2012. Viscoelastic modeling of the contact interaction between a tactile sensor and an atrial tissue. *IEEE Transactions on Biomedical Engineering* 59 (6):1727–1738.

Shen, Z. L. L., H. Kahn, R. Ballarin and S. J. Eppell. 2011. Viscoelastic properties of isolated collagen fibrils. *Biophysical Journal* 100 (12):3008–3015.

Shepherd, D. E. T. and B. B. Seedhom. 1999. Thickness of human articular cartilage in joints of the lower limb. *Annals of the Rheumatic Diseases* 58 (1):27–34.

Shiina, T. 2013. JSUM ultrasound elastography practice guidelines: Basics and terminology. *Journal of Medical Ultrasonics* 40 (4):309–323.

Shinohara, M., K. Sabra, J. L. Gennisson, M. Fink and M. Tanter. 2010. Real-time visualization of muscle stiffness distribution with ultrasound shear wave imaging during muscle contraction. *Muscle & Nerve* 42 (3):438–441.

Silver-Thorn, M. B. 1999. *In vivo* indentation of lower extremity limb soft tissues. *IEEE Transactions on Rehabilitation Engineering* 7 (3):268–277.

Simha, N. K., C. S. Carlson and J. L. Lewis. 2004. Evaluation of fracture toughness of cartilage by micropenetration. *Journal of Materials Science – Materials in Medicine* 15 (5):631–639.

Simha, N. K., H. Jin, M. L. Hall, S. Chiravarambath and J. L. Lewis. 2007. Effect of indenter size on elastic modulus of cartilage measured by indentation. *Journal of Biomechanical Engineering – Transactions of the ASME* 129 (5): 767–775.

Simon, M., J. Guo, S. Papazoglou, H. Scholand-Engler, C. Erdmann, U. Melchert, M. Bonsanto et al. 2013. Non-invasive characterization of intracranial tumors by magnetic resonance elastography. *New Journal of Physics* 15 (8):085024.

Sinkus, R., J. L. Daire, V. Vilgrain and B. E. Van Beers. 2012. Elasticity imaging via MRI: Basics, overcoming the waveguide limit, and clinical liver results. *Current Medical Imaging Reviews* 8 (1):56–63.

Sinkus, R., M. Tanter, S. Catheline, J. Lorenzen, C. Kuhl, E. Sondermann and M. Fink. 2005. Imaging anisotropic and viscous properties of breast tissue by magnetic resonance-elastography. *Magnetic Resonance in Medicine* 53 (2):372–387.

Siu, P. M., B. T. Tam, D. H. Chow, J. Y. Guo, Y. P. Huang, Y. P. Zheng and S. H. Wong. 2010. Immediate effects of 2 different whole-body vibration frequencies on muscle peak torque and stiffness. *Archives of Physical Medicine and Rehabilitation* 91 (10):1608–1615.

Soltz, M. A. and G. A. Ateshian. 1998. Experimental verification and theoretical prediction of cartilage interstitial fluid pressurization at an impermeable contact interface in confined compression. *Journal of Biomechanics* 31 (10):927–934.

Souchon, R., R. Salomir, O. Beuf, L. Milot, D. Grenier, D. Lyonnet, J. Y. Chapelon and O. Rouviere. 2008. Transient MR elastography (t-MRE) using ultrasound radiation force: Theory, safety, and initial experiments in vitro. *Magnetic Resonance in Medicine* 60 (4):871–881.

Steiger, H. J., R. Aaslid, S. Keller and H. J. Reulen. 1989. Strength, elasticity and viscoelastic properties of cerebral aneurysms. *Heart Vessels* 5 (1):41–46.

Suh, J. K. F., I. Youn and F. H. Fu. 2001. An in situ calibration of an ultrasound transducer: A potential application for an ultrasonic indentation test of articular cartilage. *Journal of Biomechanics* 34 (10):1347–1353.

Sun, C. R., B. Standish and V. X. D. Yang. 2011. Optical coherence elastography: Current status and future applications. *Journal of Biomedical Optics* 16 (4):043001.

Suresh, S. 2007. Biomechanics and biophysics of cancer cells. *Acta Biomaterialia* 3 (4):413–438.

Sutton-Tyrrell, K., S. S. Najjar, R. M. Boudreau, L. Venkitachalam, V. Kupelian, E. M. Simonsick, R. Havlik et al. 2005. Elevated aortic pulse wave velocity, a marker of arterial stiffness, predicts cardiovascular events in well-functioning older adults. *Circulation* 111 (25):3384–3390.

Swiatkowska-Freund, M. and K. Preis. 2011. Elastography of the uterine cervix: Implications for success of induction of labor. *Ultrasound in Obstetrics & Gynecology* 38 (1):52–56.

Takei, M., H. Shiraiwa, S. Omata, N. Motooka, K. Mitamura, T. Horie, T. Ookubo and S. Sawada. 2004. A new tactile skin sensor for measuring skin hardness in patients with systemic sclerosis and autoimmune Raynaud's phenomenon. *Journal of International Medical Research* 32 (2):222–231.

Tanaka, E., K. Hanaoka, T. van Eijden, M. Tanaka, M. Watanabe, M. Nishi, N. Kawai, H. Murata, T. Hamada and K. Tanne. 2003. Dynamic shear properties of the temporomandibular joint disc. *Journal of Dental Research* 82 (3):228–231.

Tanter, M., J. Bercoff, A. Athanasiou, T. Deffieux, J. L. Gennisson, G. Montaldo, M. Muller, A. Tardivon and M. Fink. 2008. Quantitative assessment of breast lesion viscoelasticity: Initial clinical results using supersonic shear imaging. *Ultrasound in Medicine and Biology* 34 (9):1373–1386.

Tanter, M., D. Touboul, J. L. Gennisson, J. Bercoff and M. Fink. 2009. High-resolution quantitative imaging of cornea elasticity using supersonic shear imaging. *IEEE Transactions on Medical Imaging* 28 (12):1881–1893.

Tepic, S., T. Macirowski and R. W. Mann. 1983. Mechanical properties of articular cartilage elucidated by osmotic loading and ultrasound. *Proceedings of the National Academy of Sciences of the United States of America – Biological Sciences* 80 (11):3331–3333.

Theret, D. P., M. J. Levesque, M. Sato, R. M. Nerem and L. T. Wheeler. 1988. The application of a homogeneous half-space model in the analysis of endothelial cell micropipette measurements. *Journal of Biomechanical Engineering – Transactions of the ASME* 110 (3):190–199.

Tian, L., M. W. L. Ko, L. K. Wang, J. Y. Zhang, T. J. Li, Y. F. Huang and Y. P. Zheng. 2014. Assessment of ocular biomechanics using dynamic ultra high-speed Scheimpflug imaging in keratoconic and normal eyes. *Journal of Refractive Surgery* 30 (11):785–U99.

Timoshenko, S. and J. N. Goodier. 1951. Axially symmetrical stress distribution in a solid of revolution. In *Theory of Elasticity*, eds. S. Timoshenko and J. N. Goodier. New York: McGraw-Hill.

Tonuk, E. and M. B. Silver-Thorn. 2003. Nonlinear elastic material property estimation of lower extremity residual limb tissues. *IEEE Transactions on Neural Systems and Rehabilitation Engineering* 11 (1):43–53.

Torr, G. R. 1984. The acoustic radiation force. *American Journal of Physics* 52 (5):402–408.

Toyras, J., M. S. Laasanen, S. Saarakkala, M. J. Lammi, J. Rieppo, J. Kurkijarvi, R. Lappalainen and J. S. Jurvelin. 2003. Speed of sound in normal and degenerated bovine articular cartilage. *Ultrasound in Medicine and Biology* 29 (3):447–454.

Toyras, J., J. Rieppo, M. T. Nieminen, H. J. Helminen and J. S. Jurvelin. 1999. Characterization of enzymatically induced degradation of articular cartilage using high frequency ultrasound. *Physics in Medicine and Biology* 44 (11):2723–2733.

Tse, Z. T. H., H. Janssen, A. Hamed, M. Ristic, I. Young and M. Lamperth. 2009. Magnetic resonance elastography hardware design: A survey. *Proceedings of the Institution of Mechanical Engineers Part H – Journal of Engineering in Medicine* 223 (H4):497–514.

Turner, C. H., J. Rho, Y. Takano, T. Y. Tsui and G. M. Pharr. 1999. The elastic properties of trabecular and cortical bone tissues are similar: Results from two microscopic measurement techniques. *Journal of Biomechanics* 32 (4):437–441.

Uffmann, K., S. Maderwald, W. Ajaj, C. G. Galban, S. Mateiescu, H. H. Quick and M. E. Ladd. 2004. In vivo elasticity measurements of extremity skeletal muscle with MR elastography. *NMR in Biomedicine* 17 (4):181–190.

Urban, M. W., A. Alizad, W. Aquino, J. F. Greenleaf and M. Fatemi. 2011. A review of vibro-acoustography and its applications in medicine. *Current Medical Imaging Reviews* 7 (4):350–359.

Urban, M. W., S. G. Chen and M. Fatemi. 2012. A review of shearwave dispersion ultrasound vibrometry (SDUV) and its applications. *Current Medical Imaging Reviews* 8 (1):27–36.

Urban, M. W., S. G. Chen and J. F. Greenleaf. 2009. Error in estimates of tissue material properties from shear wave dispersion ultrasound vibrometry. *IEEE Transactions on Ultrasonics Ferroelectrics and Frequency Control* 56 (4):748–758.

Urban, M. W., G. T. Silva, M. Fatemi and J. F. Greenleaf. 2006. Multifrequency vibro-acoustography. *IEEE Transactions on Medical Imaging* 25 (10):1284–1295.

Van Vliet, K. J. 2011. Instrument and experiment. In *Handbook of Nanoindentation with Biological Applications*, ed. M. L. Oyen. Singapore: Pan Stanford, Chapter 3.

VanLandingham, M. R. 2003. Review of instrumented indentation. *Journal of Research of the National Institute of Standards and Technology* 108 (4):249–265.

Vannah, W. M. and D. S. Childress. 1996. Indentor tests and finite element modeling of bulk muscular tissue *in vivo*. *Journal of Rehabilitation Research and Development* 33 (3):239–252.

Vannah, W. M., D. M. Drvaric, J. A. Hastings, J. A. Strand and D. M. Harning. 1999. A method of residual limb stiffness distribution measurement. *Journal of Rehabilitation Research and Development* 36 (1):1–7.

Vappou, J., C. Maleke and E. E. Konofagou. 2009. Quantitative viscoelastic parameters measured by harmonic motion imaging. *Physics in Medicine and Biology* 54 (11):3579–3594.

Varghese, T. 2009. Quasi-static ultrasound elastography. *Ultrasound Clinics* 4 (3):323–328.

Varghese, T. and J. Ophir. 1997. A theoretical framework for performance characterization of elastography: The strain filter. *IEEE Transactions on Ultrasonics Ferroelectrics and Frequency Control* 44 (1):164–172.

Varghese, T., J. Ophir, E. Konofagou, F. Kallel and R. Righetti. 2001. Tradeoffs in elastographic imaging. *Ultrasonic Imaging* 23 (4):216–248.

Varghese, T. and H. R. Shi. 2004. Elastographic imaging of thermal lesions in liver in-vivo using diaphragmatic stimuli. *Ultrasonic Imaging* 26 (1):18–28.

Venkatesh, S. K. and R. L. Ehman. 2014. Magnetic resonance elastography of liver. *Magnetic Resonance Imaging Clinics of North America* 22 (3):433–446.

Vergari, C., P. Rouch, G. Dubois, D. Bonneau, J. Dubousset, M. Tanter, J. L. Gennisson and W. Skalli. 2014. Non-invasive biomechanical characterization of intervertebral discs by shear wave ultrasound elastography: A feasibility study. *European Radiology* 24 (12):3210–3216.

Verhaegen, Pdhm, M. B. A. van der Wal, E. Middelkoop and P. P. M. van Zuijlen. 2011. Objective scar assessment tools: A clinimetric appraisal. *Plastic and Reconstructive Surgery* 127 (4):1561–1570.

Viren, T., S. Saarakkala, E. Kaleva, H. J. Nieminen, J. S. Jurvelin and J. Toyras. 2009. Minimally invasive ultrasound method for intra-articular diagnostics of cartilage degeneration. *Ultrasound in Medicine and Biology* 35 (9):1546–1554.

Viren, T., S. Saarakkala, V. Tiitu, J. Puhakka, I. Kiviranta, J. S. Jurvelin and J. Toyras. 2011. Ultrasound evaluation of mechanical injury of bovine knee articular cartilage under arthroscopic control. *IEEE Transactions on Ultrasonics Ferroelectrics and Frequency Control* 58 (1):148–155.

Viren, T., S. Saarakkala, J. S. Jurvelin, H. J. Pulkkinen, V. Tiitu, P. Valonen, I. Kiviranta, M. J. Lammi and J. Toyras. 2010. Quantitative evaluation of spontaneously and surgically repaired rabbit articular cartilage using intra-articular ultrasound method in situ. *Ultrasound in Medicine & Biology* 36 (5):833–839.

von Bally, G., B. Kemper and S. Knoche. 2002. Dynamic holographic endoscopy: New perspectives in minimally invasive diagnostics. *Medical Laser Application* 17 (1):59–64.

Vuskovic, V. 2001. Device for in-vivo measurement of mechanical properties of internal human soft tissues. PhD thesis, ETH, Swiss Federal Institutes of Technology in Zurich, Switzerland.

Wang, C. Z. 2011. Development of a vibro-ultrasound method for muscle stiffness measurement in vivo. PhD thesis, Department of Health Technology and Informatics, Hong Kong Polytechnic University, Hung Hom, Hong Kong.

Wang, C. Z., J. Y. Guo and Y. P. Zheng. August 2010a. Muscle elasticity measurement using ultrasound at isometric step contraction. Paper read at *Sixth World Congress of Biomechanics (WCB)*, Singapore.

Wang, C. Z., T. J. Li and Y. P. Zheng. 2014. Shear modulus estimation on vastus intermedius of elderly and young females over the entire range of isometric contraction. *PLoS ONE* 9 (7):e101769.

Wang, C. Z. and Y. P. Zheng. 2009. Local arterial stiffness measurement using a high frame rate ultrasound system. In *Eighth International Conference on the Ultrasonic Measurement and Imaging of Tissue Elasticity*, Vlissingen, Zeeland, the Netherlands.

Wang, C. Z. and Y. P. Zheng. 2011. Development of a vibro-ultrasound method for skeleton muscle stiffness assessment under high intensity step isometric contraction levels. In *10th International Tissue Elasticity Conference*, Arlington, TX.

Wang, J. H. C. and B. P. Thampatty. 2006. An introductory review of cell mechanobiology. *Biomechanics and Modeling in Mechanobiology* 5 (1):1–16.

Wang, L. K., J. Y. Zhang, Y. Hon, T. J. Li, Y. P. Huang, A. Lam and Y. P. Zheng. 2013. Dynamic optical coherence tomography (OCT) based air jet indentation to estimate corneal elastic property. In *The 12th International Tissue Elasticity Conference*, Lingfield, U.K.

Wang, L. K., J. Y. Zhang, L. Tian, M. W. L. Ko, Y. F. Huang and Y. P. Zheng. 2015. OCT based air jet indentation for corneal biomechanical assessment. *Optics and Precision Engineering* 23 (2):325–333.

Wang, Q. 2007. Ultrasound monitoring of transient and inhomogeneous swelling of articular cartilage. PhD thesis, Health Technology and Informatics, Hong Kong Polytechnic University, Hung Hom, Hong Kong.

Wang, Q., Y. P. Zheng, H. J. Niu and A. F. T. Mak. 2007. Extraction of mechanical properties of articular cartilage from osmotic swelling behavior monitored using high frequency ultrasound. *Journal of Biomechanical Engineering – Transactions of the ASME* 129 (3):413–422.

Wang, S. Z., Y. P. Huang, S. Saarakkala and Y. P. Zheng. 2010b. Quantitative assessment of articular cartilage with morphologic, acoustic and mechanical properties obtained using high frequency ultrasound. *Ultrasound in Medicine & Biology* 36 (3):512–527.

Wang, S. Z., Y. P. Huang, Q. Wang, Y. P. Zheng and Y. H. He. 2010c. Assessment of depth and degeneration dependences of articular cartilage refractive index using optical coherence tomography *in vitro*. *Connective Tissue Research* 51 (1):36–47.

Wanninayake, I. B., L. D. Seneviratne and K. Althoefer with IEEE. 2012. Novel indentation depth measuring system for stiffness characterization in soft tissue palpation. In *2012 IEEE International Conference on Robotics and Automation*. New York: IEEE.

Watanabe, T., S. Omata, J. Z. Lee and C. E. Constantinou. 1997. Comparative analysis of bladder wall compliance based on cystometry and biosensor measurements during the micturition cycle of the rat. *Neurourology and Urodynamics* 16 (6):567–581.

Waters, N. E. 1965. Indentation of thin rubber sheets by cylindrical indentors. *British Journal of Applied Physics* 16 (9):1387–1392.

Wells, P. N. T. and H. D. Liang. 2011. Medical ultrasound: Imaging of soft tissue strain and elasticity. *Journal of the Royal Society Interface* 8 (64):1521–1549.

Wilde, G. S., H. J. Burd and S. J. Judge. 2012. Shear modulus data for the human lens determined from a spinning lens test. *Experimental Eye Research* 97 (1):36–48.

Wineman, A. 2009. Nonlinear viscoelastic solids – A review. *Mathematics and Mechanics of Solids* 14 (3):300–366.

Wollensak, G. and E. Spoerl. 2004. Biomechanical characteristics of retina. *Retina-the Journal of Retinal and Vitreous Diseases* 24 (6):967–970.

Wong, G. L. H., V. W. S. Wong, P. C. L. Choi, A. W. H. Chan, A. M. L. Chim, K. K. L. Yiu, S. H. T. Chu, F. K. L. Chan, J. J. Y. Sung and H. L. Y. Chan. 2011. On-treatment monitoring of liver fibrosis with transient elastography in chronic hepatitis B patients. *Antiviral Therapy* 16 (2):165–172.

Wu, J. Z., R. G. Dong and A. W. Schopper. 2004a. Analysis of effects of friction on the deformation behavior of soft tissues in unconfined compression tests. *Journal of Biomechanics* 37 (1):147–155.

Wu, J. Z., R. G. Dong, W. P. Smutz and A. W. Schopper. 2003. Nonlinear and viscoelastic characteristics of skin under compression: Experiment and analysis. *Bio-Medical Materials and Engineering* 13 (4):373–385.

Wu, Z., K. Hoyt, D. J. Rubens and K. J. Parker. 2006. Sonoelastographic imaging of interference patterns for estimation of shear velocity distribution in biomaterials. *Journal of the Acoustical Society of America* 120 (1):535–545.

Wu, Z., L. S. Taylor, D. J. Rubens and K. J. Parker. 2004b. Sonoelastographic imaging of interference patterns for estimation of the shear velocity of homogeneous biomaterials. *Physics in Medicine and Biology* 49 (6):911–922.

Xie, J. P., J. B. Zhou and Y. C. Fung. 1995. Bending of blood vessel wall: Stress-strain laws of the intima-media and adventitial layers. *Journal of Biomechanical Engineering – Transactions of the ASME* 117 (1):136–145.

Xiong, S., R. S. Goonetilleke, C. P. Witana and W. D. A. S. Rodrigo. 2010. An indentation apparatus for evaluating discomfort and pain thresholds in conjunction with mechanical properties of foot tissue *in vivo*. *Journal of Rehabilitation Research and Development* 47 (7):629–641.

Xu, L., Y. Lin, J. C. Han, Z. N. Xi, H. Shen and P. Y. Gao. 2007. Magnetic resonance elastography of brain tumors: Preliminary results. *Acta Radiologica* 48 (3):327–330.

Xydeas, T., K. Siegmann, R. Sinkus, U. Krainick-Strobel, S. Miller and C. D. Claussen. 2005. Magnetic resonance elastography of the breast – Correlation of signal intensity data with viscoelastic properties. *Investigative Radiology* 40 (7):412–420.

Yamakoshi, Y., J. Sato and T. Sato. 1990. Ultrasonic imaging of internal vibration of soft tissue under forced vibration. *IEEE Transactions on Ultrasonics Ferroelectrics and Frequency Control* 37 (2):45–53.

Yanai, M., J. P. Butler, T. Suzuki, A. Kanda, M. Kurachi, H. Tashiro and H. Sasaki. 1999. Intracellular elasticity and viscosity in the body, leading, and trailing regions of locomoting neutrophils. *American Journal of Physiology – Cell Physiology* 277 (3):C432–C440.

Yang, Y., P. O. Bagnaninchi, M. Ahearne, R. K. Wang and K. K. Liu. 2007. A novel optical coherence tomography-based micro-indentation technique for mechanical characterization of hydrogels. *Journal of the Royal Society Interface* 4 (17):1169–1173.

Yao, H., M. A. Justiz, D. Flagler and W. Y. Gu. 2002. Effects of swelling pressure and hydraulic permeability on dynamic compressive behavior of lumbar annulus fibrosus. *Annals of Biomedical Engineering* 30 (10):1234–1241.

Yeh, W. C., P. C. Li, Y. M. Jeng, H. C. Hsu, P. L. Kuo, M. L. Li, P. M. Yang and P. H. Lee. 2002. Elastic modulus measurements of human liver and correlation with pathology. *Ultrasound in Medicine and Biology* 28 (4):467–474.

Yoganandan, N., S. Kumaresan and F. A. Pintar. 2000. Geometric and mechanical properties of human cervical spine ligaments. *Journal of Biomechanical Engineering – Transactions of the ASME* 122 (6):623–629.

Young, W. C. 1989. *Roak's Formulas for Stress and Strain.* New York: McGraw-Hill.

Yu, Q. L., J. B. Zhou and Y. C. Fung. 1993. Neutral axis location in bending and Young's modulus of different layers of arterial wall. *American Journal of Physiology* 265 (1):H52–H60.

Yuan, Y. H. and R. Verma. 2006. Measuring microelastic properties of stratum corneum. *Colloids and Surfaces B – Biointerfaces* 48 (1):6–12.

Zdero, R., M. Olsen, S. Elfatori, T. Skrinskas, H. Nourhosseini, C. Whyne, E. H. Schemitsch and H. von Schroeder. 2009. Linear and torsional mechanical characteristics of intact and reconstructed scapholunate ligaments. *Journal of Biomechanical Engineering – Transactions of the ASME* 131 (4):041009.

Zhai, L., M. L. Palmeri, R. R. Bouchard, R. W. Nightingale and K. R. Nightingale. 2008. An integrated indenter-ARFI imaging system for tissue stiffness quantification. *Ultrasonic Imaging* 30 (2):95–111.

Zhang, M., Y. P. Zheng and A. F. T. Mak. 1997. Estimating the effective Young's modulus of soft tissues from indentation tests – Nonlinear finite element analysis of effects of friction and large deformation. *Medical Engineering & Physics* 19 (6):512–517.

Zheng, Y. P. 1997. Development of an ultrasound indentation system for biomechanical properties assessment of limb tissues in vivo. PhD thesis, Rehabiliation Engineering Center, Hong Kong Polytechnic University, Hung Hom, Hong Kong.

Zheng, Y. P., S. L. Bridal, J. Shi, A. Saied, M. H. Lu, B. Jaffre, A. F. T. Mak, and P. Laugier. 2004. High resolution ultrasound elastomicroscopy imaging of soft tissues: system development and feasibility. *Physics in Medicine and Biology* 49 (17):3925-3938.

Zheng, Y. P., J. G. Chen and H. Y. Ling. 2011. Development of an ultrasound platform for the evaluation of plantar soft tissue properties: A feasibility study on silicone phantom feet. *Instrumentation Science & Technology* 39 (3):248–260.

Zheng, Y. P., A. P. C. Choi, H. Y. Ling, and Y. P. Huang. 2009. Simultaneous estimation of Poisson's ratio and Young's modulus using a single indentation: a finite element study. *Measurement Science & Technology* 20 (4):045706.

Zheng, Y. P., Y. K. C. Choi, K. Wong, S. Chan and A. F. T. Mak. 2000a. Biomechanical assessment of plantar foot tissue in diabetic patients using an ultrasound indentation system. *Ultrasound in Medicine and Biology* 26 (3):451–456.

Zheng, Y. P., C. X. Ding, J. Bai, A. F. T. Mak, and L. Qin. 2001. Measurement of the layered compressive properties of trypsin-treated articular cartilage: An ultrasound investigation. *Medical & Biological Engineering & Computing* 39 (5):534–541.

Zheng, Y. P. and Y. P. Huang. 2008. More intrinsic parameters should be used in assessing degeneration of articular cartilage with quantitative ultrasound. *Arthritis Research & Therapy* 10 (6):125.

Zheng, Y. P., Y. P. Huang, Y. P. Zhu, M. Wong, J. F. He and Z. M. Huang. 2012. Development of a foot scanner for assessing the mechanical properties of plantar soft tissues under different bodyweight loading in standing. *Medical Engineering & Physics* 34 (4):506–511.

Zheng, Y. P., S. F. Leung and A. F. T. Mak. 2000b. Assessment of neck tissue fibrosis using an ultrasound palpation system: A feasibility study. *Medical & Biological Engineering & Computing* 38 (5):497–502.

Zheng, Y. P., R. G. Maev and I. Y. Solodov. 1999a. Nonlinear acoustic applications for material characterization: A review. *Canadian Journal of Physics* 77 (12):927–967.

Zheng, Y. P. and A. F. T. Mak. 1996. An ultrasound indentation system for biomechanical properties assessment of soft tissues *in-vivo*. *IEEE Transactions on Biomedical Engineering* 43 (9):912–918.

Zheng, Y. P. and A. F. T. Mak. 1999. Extraction of quasi-linear viscoelastic parameters for lower limb soft tissues from manual indentation experiment. *Journal of Biomechanical Engineering – Transactions of the ASME* 121 (3):330–339.

Zheng, Y. P., A. F. T. Mak, K. P. Lau, and L. Qin. 2002. An ultrasonic measurement for in vitro depth-dependent equilibrium strains of articular cartilage in compression. *Physics in Medicine and Biology* 47 (17):3165–3180.

Zheng, Y. P., A. F. T. Mak and B. Lue. 1999b. Objective assessment of limb tissue elasticity: Development of a manual indentation procedure. *Journal of Rehabilitation Research and Development* 36 (2):71–85.

Zheng, Y. P., H. J. Niu, F. T. A. Mak, and Y. P. Huang. 2005. Ultrasonic measurement of depth-dependent transient behaviors of articular cartilage under compression. *Journal of Biomechanics* 38 (9):1830–1837.

Zheng, Y. P., L. K. Wang, T. J. Li and Y. Y. Wang. 2014. An OCT-based air suction-indentation probe for tissue elasticity measurement. Paper read at *SPIE, Optical Elastography and Tissue Biomechanics,* San Francisco, CA.

Zheng, Y. P., T. M. Mak, Z. M. Huang, C. W. J. Cheung, Y. J. Zhou and J. F. He. August 1–6, 2010. Liver fibrosis assessment using transient elastography guided with real-time B-mode ultrasound imaging. In *Sixth World Congress of Biomechanics (WCB 2010)*, Singapore, eds. C. T. Lim and J. C. H. Goh, Berlin, Germany: Springer.

Ziegert, J. C. and J. L. Lewis. 1978. In-vivo mechanical properties of soft tissues covering bony prominences. *Journal of Biomechanical Engineering – Transactions of the ASME* 100 (4):194–201.

Ziol, M., A. Handra-Luca, A. Kettaneh, C. Christidis, F. Mal, F. Kazemi, V. de Ledinghen et al. 2005. Noninvasive assessment of liver fibrosis by measurement of stiffness in patients with chronic hepatitis C. *Hepatology* 41 (1):48–54.

Index